Evolution's Eye

Science and Cultural Theory

A Series Edited by Barbara Herrnstein Smith

and E. Roy Weintraub

Evolution's Eye

A Systems View of the Biology-Culture Divide

Susan Oyama

DUKE UNIVERSITY PRESS   Durham and London 2000

Contents

Acknowledgments

The following people were of special help in writing this book. By offering comments on the pieces when they were originally being written, advice and other aid at varying points in the preparation of the book itself, or both, they demonstrated once again the value of critical and generous colleagues. My deep thanks go to Pat Bateson, Ben Bradley, Linnda Caporael, Richard Francis, Eli Gerson, Peter Godfrey-Smith, Russell Gray, Paul Griffiths, Megan Gunnar, Jack Hailman, Mae-Wan Ho, Tim Johnston, Evelyn Fox Keller, William Kessen, Philip Kitcher, Richard Lewontin, Lenny Moss, Katherine Nelson, Jeff Ricker, Barbara Herrnstein Smith, Elliott Sober, Don Symons, Peter Taylor, Rasmus Winther, and Cor van der Weele. Miriam Angress, Pam Morrison, and Reynolds Smith of Duke University Press were responsive and supportive; what more could any author ask?

Introduction

This book is about an approach to development and evolution that provides a novel way of analyzing these two biological processes and their interrelationships. It also allows us to bring them together quite differently, by means of the concept of a developmental system: a heterogeneous and causally complex mix of interacting entities and influences that produces the life cycle of an organism. The system includes the changing organism itself, because an organism contributes to its own future, but it encompasses much else as well.

Although the notion of a developmental system alters our understanding of development and evolution, it also has significant implications for a multitude of broader issues, scientific and otherwise. This is because contemporary versions of the age-old distinctions between biology and culture, between nature and nurture, are grounded in largely unquestioned assumptions about how the processes of development and evolution are, or are not, conjoined.

Developmental Systems: DST, DSA, DSP?

In my earlier writings I have spoken primarily for myself, however much I have drawn on the work of others. My later writings refer to a sometimes shadowy band of "developmental systems theorists," and some readers have wondered whether these constitute a school of sorts, and if so, what its shape and extent are (not to mention "how to buy some of it and take it home," as Barbara Herrnstein Smith [1988:180] says in a slightly different context). That the question of group identity should arise has its pleasing aspect. It seems that scholars from a variety of disciplines are finding in developmental systems writings a fresh way of addressing a wide range of issues and feel the need for a convenient label. Occasionally I have used the term *developmental systems theory* (DST) to describe this approach, and I will continue to use it here.[1] It is short and snappy, already in use, and convenient for indicating a rough grouping of alternative conceptualizations and critiques. I have tended to employ the term with qualifications, however, indicating that if a theory is taken to be only a hypothesis-generating machine, then perhaps *perspective* or *approach* would be preferable. What I have in mind is broader: a kind of conceptual background that serves to orient more specific empirical and theoretical endeavors. It can also allow different interpretations of existing work. (See van Gelder and Port 1995 for a discussion of this framing or orienting function of approaches to cognition.)

The need for a term cannot be ignored, but any act of individuation involves the marking of commonalities and dissimilarities in order to pick something out from the background. The drawing of such outlines can be a delicate matter. Reciting a creed or compiling lists of insiders risks essentializing a more or less loose assemblage in inappropriate ways. Nor is this the place for a history or an analysis of related approaches.[2] I will, however, sketch a number of key ideas and methods that characterize the work presented in this book. Fuller presentations are found primarily in the chapters of part I; to varying degrees these features also characterize the writings of the colleagues with whom I am most closely associated.[3] Although we don't always agree on technical details or on the precise form and limits of necessary theoretical reworkings, I think we share a commitment to a view of development as interactive emergence over time and to exploring the implications of taking this gradual construction seriously. Following the ideas and methods below is a list of work that was especially important for the development of my thinking, along with some remarks on research and metaphor.

Key Ideas and Methodological Strategies

1. *Parity of reasoning.* A prime characteristic of my method is an insistence on applying certain modes of reasoning consistently and rigorously, even in areas in which they are not customarily used. Parity-of-reasoning arguments (namely, what's sauce for the goose . . .) are especially useful in revealing hidden inequalities and questionable assumptions. In chap-

ters 1 and 2, for instance, I show how the reasons for calling features of organisms "genetic" or "environmental" tend to be used asymmetrically. These analyses are developed in subsequent chapters (especially 6) to demonstrate how genes are frequently given causal priority even when similar reasoning could be used for other causes. In such cases, what is presented as a *justification* for giving the gene special status is actually a *consequence of already having done so.* Undoing these conventional asymmetries highlights, among other things, point 2.

2. *The developmental and evolutionary interdependence of organism and environment.* I replace conventional wisdom's "interaction" between separate and independently defined organisms and environments with "constructivist interaction." This is not to be confused with the "inter-action term" in some statistical analyses; that is why "constructivist" is important. In constructivist interaction organisms and their environments define the relevant aspects of, and can affect, each other. This is called the "interpenetration" of organism and environment by, for example, Lewontin (1982; see also Lewontin, Rose, and Kamin 1984). Seeing life in these terms rather than following the traditional cleavages between nature and culture, body and mind, permits different and fruitful ways of conceiving biology, psychology, and society, as well as different relations among the disciplines (see Ingold 1991). Such a view suggests point 3.

3. *A shift from "genes and environment" to a multiplicity of entities, influences, and environments.* One benefit of thinking in terms of multiple systems, interconnected and measurable on more than one scale of time and magnitude, is that the stunningly oversimplified distinction between genes and environment resolves into a heterogeneous and equally stunning array of processes, entities, and environments—chemical and mechanical, micro- and macroscopic, social and geological—of the sort outlined in chapter 3. Evolution's eye widens, and so do ours. Hence my preference in the following chapters for systemic interaction over one-way causal arrows, for mutually influencing levels rather than the collapsed levels that often go with unidirectional causation. The shift from genes-and-environment to multilevel systems is coupled with the next point.[4]

4. *A shift from single to multiple scales.* Looking through developmental systems lenses means, among other things, looking hard and repeat-

edly in a variety of directions and with both near and distant focuses. Movement among scales, both of magnitude and time, is important: from interactions of molecules inside cells to those between persons, from the brief periods involved in the action of a hormone in the nervous system to changing relations among conspecifics over the life span, from the short-term dynamics of a population of organisms in a habitat to the slow procession of generations through evolutionary time.[5]

Moving among scales in this way not only enables us to see *more,* it also gives us a more acute sense of the many relations among these scales. The outcome of an aggressive encounter between animals, for instance, can lead to physiological changes, including changes in gene activity (see Gottlieb 1970, 1992, 1997, on bidirectional influences between structure and function; Lehrman 1962). Similarly, rapid processes can influence, and be influenced by, much slower ones, as might happen if gradual climatic change altered the kind of food available to a creature. The animal's digestive processes might adapt (with or without accompanying genetic change), and this might eventuate in a different foraging range, which could then lead to other developmental and evolutionary changes (see P. P. G. Bateson 1988a; Gray 1988; Johnston and Gottlieb 1990; Lewontin 1982; Sterelny, Smith, and Dickison 1996).

There is still a place for the "provisional single-mindedness" mentioned in chapter 3, the tactical decision to focus on one level for a particular purpose. It is important to recognize that such an approach temporarily delimits the context for the investigation; eventually the questions and findings must be recontextualized. There is, after all, no research that does not limit the scope of entities and variables studied. Items 1 through 4 lead ineluctably (for me, at least) to item 5.

5. *Extending heredity.* I have taken two kinds of critiques, of innateness in individuals and of the relations between evolving populations and their niches, and connected them by means of a broader notion of heredity than has been countenanced in contemporary science. A good part of my writing, the present volume included, involves the working out of that connection and its many ramifications both inside and outside biology.

The earlier work of the American comparative psychologists and English ethologists who are mentioned below in "Resources" is probably the most widely known for questioning internalist conceptions of innateness and instinct. Lewontin (e.g., 1982) wrote on these topics as well, but also mounted a powerful attack on the externalist conception of natural selection. He challenged the belief that the environment presents organisms with already defined problems and opportunities, to which the organisms must then adapt. This is the evolutionary analogue of the idea that developmental "information" exists in the genes and in the environment rather than being generated in constructivist interaction, and Lewontin's "interpenetration of organism and environment" can, among other things, be seen as the evolutionary analogue of the mutual dependence and mutual construction of organism and developmental environment in ontogeny (i.e., the development of individual organisms) described in chapters 1, 4, and 5.

The debates on innateness and evolutionary adaptation were significant in themselves; indeed, they are still in progress. Until development and evolution were related in a different way, however, each critique remained in a certain sense incomplete, and the conceptual difficulties at which they were directed were bound to be repeatedly regenerated, in large part by prevailing notions of heredity. It is, in fact, mainly the *synthesis* of constructivist interactionist visions of ontogeny and of phylogeny that has attracted the attention of philosophers of biology (Godfrey-Smith 1996, 1999; Schaffner 1998; Sterelny, Smith, and Dickison 1996; Sterelny and Griffiths 1999; Winther 1996).

Because thinking about heredity is tightly tied to the genecentric notions of programmed development described in a number of the following chapters, a reworked heredity goes hand in hand with the next point.

6. *A shift from central control to interactive, distributed regulation.* The aptness of constructivist interaction is not restricted to the level of the organism. There is not, in fact, only one such level; all organisms contain multiple levels, and all participate in wider ecological complexes. Applied to genetic processes, the ideas of mutual definition, dependence, and influence reveal the inadequacy of usual accounts of genetically controlled development. As the phrase *developmental system* suggests, "control" is instead multiple and mobile, distributed and systemic.

Thinking in terms of developmental systems invites certain kinds of inquiry and provides a conceptual frame within which to interpret the answers. Inferential bobbles of the sort cited in chapter 3, for instance, are rendered less likely. Researchers' attention can be directed in novel ways, toward often-overlooked "background" factors. Once we stop relying on a central agency to bring these factors together, furthermore, a new realm of questions opens up, many of which are unlikely to be asked otherwise, about when, how, and how reliably these factors do in fact come together (see below, and chapters 4 and 6). To claim that the genes contain already formed programs, representations, "information," or other prime movers is not only mistaken, it is to miss the contextualized richness of these processes. To capture these processes, we need item 7.

7. *A shift from transmission to continuous construction and transformation.* We move, then, from hereditary transmission of traits, or coded representations of traits, to the continuous developmental construction and transformation of organisms and their worlds in repeating life cycles. As chapters 1, 3, 4, and 5 indicate, a changed understanding of development alters our understanding of evolution. Thus we have point 8.

8. *Theoretical extension and unification.* The first seven points should convey the senses of *system* and *interaction* intended here (Oyama, in press). In addition to being analytically valuable, the systematic evenhandedness of DST's parity-of-reasoning arguments can also be used constructively. One can retheorize factors customarily relegated to secondary status by more genecentric theorists, for instance, and treat them as full-fledged participants in an explanatory scheme. The insistence on symmetry thus has an extremely important function: the theoretical extension of rules and definitions into new domains. This is one of the major ways that theory is strengthened and unified. If a criterion for inheritance can apply to other developmental factors besides genes or cytoplasm, this changes the range of things that can be inherited, and so transforms the meaning of inheritance itself.

The consequences of this change are striking. Attempts to deal with "biology" and "culture" have usually involved positing two channels of information, and even then much is left out (see chapters 1 and 11). In place of discrete channels, I give a single framework for speaking about heredity: the changing system of "interactants" and resources that make and remake an organism and its environments throughout the life cycle. Whether inheritance is construed in terms of *resources* for the development of individuals or of *difference makers* in populations (see chapter 4), there is no principled way of restricting it to genes or germ cells. This makes it impossible to see biological and cultural evolution as separate (even if somehow "interacting") processes.

Ironically, what may strike newcomers and skeptics as excessive complexity ("Too many things are inherited") turns out to be the very simplicity and elegance that has traditionally been considered the hallmark of good theory.

Resources

As I noted earlier, people working in DST come from a variety of backgrounds; certainly they draw on a diversity of sources. I would not presume to enumerate the influences on these colleagues, but because it may help those who are new to this perspective to learn a bit about where it comes from, I will mention several bodies of writing that have been especially rich resources for my own work. While not a family tree, this roster fulfills some of the functions of a scholarly genealogy by identifying key sources and affinities. It is frequently the case in academia (and outside it, too) that one has some leeway in naming one's ancestors, even one's cousins. The following catalogue is personal, incomplete, and generalized. Not surprisingly, it includes people who have been alert to the disorder and complication lurking behind the deceptive tidiness of certain widely accepted categories and oversimplified models, as well as to the largely unrecognized order in many developmental exchanges between organisms and their milieus. This kind of regularity is often missed by researchers too ready to assume "randomness" of environmental contingencies.

Tributaries include the comparative psychologists and psychobiologists T. C. Schneirla (1966, 1972), Daniel Lehrman (1953, 1970), Peter Klopfer (1973), Zing-Yang Kuo (1922, 1976), Gilbert Gottlieb (work summarized in his 1997), and Timothy Johnston (1982, 1987). Especially valuable was these researchers' willingness not only to acknowledge complexity, but to work with it by following influences in more than one direction. Also significant were the English ethologists whose early debates with Lorenz (1965) over the concept of instinct helped open up the topic to critical scrutiny, just as Lehrman's did. Their interweaving of developmental and evolutionary topics has done much to bridge the gap between those two traditions (P. P. G. Bateson 1985, 1988a; Hinde 1968; Hinde and Bateson 1984). Another stream comes from the critical work of geneticist Richard Lewontin and his associates (1982; Levins and Lewontin 1985; Lewontin, Rose, and Kamin 1984); especially important were their writings on the mutual definition and construction of organisms and their environments in evolution, and on analytical techniques in biology.

In no case did these scholars' interest in evolution militate against a finely tuned sensitivity to the subtleties of developmental dynamics, and their explorations of ecological relations have been equally perceptive. Their accomplishments are a decisive demonstration that it is both possible and productive of evolutionary insight to broach the "black box" of development.

Research

The writings mentioned above are rich with examples of research on the kinds of relationships described in my section on ideas and strategies. Many others can be found in the works cited in note 3. Russell Gray (1988) brings together some especially useful references, although he notes that research on what he calls "evolutionary and ecological cascades" is not plentiful, probably because the questions have not been posed in appropriate ways. Similarly, Cor van der Weele (1999), who offers many findings on environmental factors in development, notes the relative neglect of this topic. Eva Jablonka and Marion Lamb (1995:ix) supply many citations in addition to their own findings on "epigenetic inheritance," and observe that existing examples of such phenomena tend to be overlooked.

Indeed, a "body of research," as opposed to a mass of disconnected and hard-to-find bits, comes into being through the synthetic conceptual work in which these people are engaged. What an instance of empirical work "shows" can change as the observations are placed within a conceptual framework that helps them make sense and from which other research can be generated.

Directions for research discussed in the present volume (especially in

chapters 3 and 4) include looking for ecological, behavioral, and physiological links between generations, as well as asking how intergenerational changes can be maintained, damped, or amplified. Jablonka and
Lamb (1995) take up issues of damping and amplification at the cellular
level, as does Mae-Wan Ho (1984). Links among Tinbergen's (1963) "four
whys" (discussed in chapter 3) are also promising research lodes. Richard
Levins and Richard Lewontin (1985), Patrick Bateson (1987, 1988a), and
many others named here give wonderful accounts of the ways in which
organisms construct, and are affected by, their developmental and evolutionary environments. Sometimes those causal complexes can come into
being again and again, making for the stably repeating generations that
permit the language of "gene flow." There are enough investigatory leads
here to keep numerous laboratories going indefinitely.

Metaphor and Practice

Some of the above-mentioned work is marked by considerable sensitivity
to language—to the images and similes that help make the world intelligible. There is, in fact, a body of writings that is especially illuminating
about the ways these metaphors are involved in the practice and social
uses of science. Evelyn Fox Keller (1985) has been particularly insight-
ful about the importance of metaphor in creating and maintaining genecentrism, as have Richard Doyle (1997), Jan Sapp (1987), and Cor van der
Weele (1999). Certain scholars in the "developmental constraints" tradition discussed below (Goodwin 1970; Ho 1988a; Ho and Saunders 1979;
Webster and Goodwin 1982) have published useful analyses as well.

Some readers have been drawn to what they perceive to be the ethical
implications of the idea of developmental systems. Although social and
ethical concerns inform much of my writing, as they do the other analyses just cited, and though these concerns are surely visible in my choices
of topic and metaphor, and in myriad other ways, I do not believe that
direct guidance or justification for particular moral stands is to be found
in my approach. As I point out in chapter 8, one can invoke scientific perspectives for their resonances, imagery, insights, and even inspirations
without supposing that, *in themselves,* those perspectives show the one
true way to a better world. If they did, there would be less need for the

kind of difficult and innovative work on ecologically sensitive practices that is discussed there.

To be sure, scientific metaphors involve not just ways of talking and writing, but ways of seeing and doing as well. They are implicated in the practices of research, from the initial direction of attention right on to the interpretation, promulgation of results, and application. My choice of title reflects a conviction that one cannot talk about matters of theory and practice without attending to the nuances of language.

Evolution's Eyes

Evolution's Eye is meant to be read in several ways. Among other things, it reflects my preoccupation with the relationship between looking (Who looks, and with what questions and assumptions in mind?) and seeing (How does what is seen bear the mark of the seer, even as it affects that seer?). To those engaged in evolutionary studies, however, what may come to mind on first seeing this title is the structure of the eye itself, for that organ has long been a powerful prompt to speculation about the origins of complex structures. Both the perfection of the eye and its imperfection have figured in arguments about what kind of artificer, divine or natural, could account for its existence. But it is not the eye as actual and marvelous object that provides the best entrée into the essays collected here, but rather several other eyes: the critical eye of natural selection, the narrow eye of the needle in the parable of the camel, and the first-person *I* that grammatically indicates a self. The epistemological *I* is here, too—the *I,* as well as the *eye,* of the knower. Some comments on these glosses will introduce my major themes.

SELECTING EYE, BLIND EYE

Evolution, or natural selection, with which evolution is too often identified, is frequently depicted as an agent that continuously scrutinizes organisms in order to identify those best suited to life. It engineers their improvement, producing, in fact, things such as organs of sight. Evolution's eye in this sense is a critical eye, measuring, comparing, and evaluating.

It is sometimes said that natural selection "doesn't care about" or "can't see" characters that make no difference to the organism's chances of surviving or producing viable offspring. Most notably in the present case, evolution is said to be blind to a living being's precise means of developing. It is developmental *outcomes* that are important, the argument goes, outcomes that are ultimately priced out in the currency of reproductive success. In this view the mechanisms by which those outcomes come about are rather beside the point, the natural selectionist point being conventionally defined in terms of the relative frequencies of genes in populations, not the particulars of the lives of individuals.

A fundamental topic in biology, then, namely the coming into being of living things, their growth and changes over the span of their lifetimes—in short, the cycling of their very embodied lives in their worlds—has a puzzlingly tenuous connection to dominant theories of the evolutionary process. How is this possible? Is it necessary? These are two of the problems around which these collected essays circle. Though there are interesting historical issues to be explored here, I have been occupied mainly with the interlocking habits of thought and practice that appear to condemn development to play a supporting role in the evolutionary spectacle, "black-boxed," as suggested above, if it is recognized at all.

Developmental processes are not completely marginalized in evolutionary studies. There is a fair amount of work on "developmental constraints on evolution": limits on the sorts of forms that can be produced in ontogeny, and thus the number and kinds of variants that can be presented for inspection by natural selection. This is an important and varied literature, and I share many of its authors' concerns. In general, though, I find their critical points more congenial than their solutions. I disagree on two points in particular. First, they tend to view development as "internal," whereas, for reasons most fully presented in chapter 5, I believe it makes more sense to see it as a matter of constructivist interaction between insides and outsides.[6] The conventional view of development as driven from within enforces the conventional ways of linking development to evolution, as an internal factor that can "interact" with an external natural selection. In this case the "interlocking habits of thought and practice" I study tie internalist views of ontogeny to externalist ones of selection. This in turn has prevented valid critiques from being fully assimilated, as I pointed out above.

My second point of disagreement with students of developmental constraints and kindred concepts is that insofar as they accept the definition of evolution (only and always, or at least fundamentally) as a change in the relative frequencies of alternative genetic variants across generations,[7] their protests may remain relatively easy to accommodate by footnotes and parentheses. Their arguments, however compelling, can still be treated as addenda, as lists of secondary factors "biasing" the action of selection. The standard assumption is that evolutionary relevance requires a particular kind of visibility to natural selection: Phenotypic variations—that is, variations in the characteristics and behavior of actual organisms—must be "coded by" or "caused by," or at least correlated with, genetic variants. Only in this way, it is believed, can differential survival and reproduction lead to the change in genetic frequencies now considered criterial for evolution.[8] But one can also say that this elevates *one aspect* of evolution to its very *definition*.

To adopt a thoroughgoing interactive constructivism with respect to both developmental and evolutionary processes is to treat the "interaction between development and evolution" sometimes mentioned in the constraints literature as something more than an easily betrayed platitude—a dictum that remains "true" (in its fashion) because its implication of separate ontogenetic and phylogenetic processes can be assented to without being examined. The consequences of adopting a developmental systems perspective are far-reaching. Among other things, one must embark on the quite radical reworking of development, inheritance, and evolution that is undertaken in this volume.

THE EYE OF THE NEEDLE

Linked to the first reading of evolution's eye, and so to the puzzle of the place of development in evolution, is a second one: the eye of the reproductive needle. This is sometimes referred to as the "bottleneck of reproduction": the narrowing of the life cycle between successive generations of certain organisms to a single cell (Bonner 1974). This one-celled stage is not universal, as anyone knows who has watched a runner or a leaf grow into a flourishing plant. Consider, however, organisms that do begin a new life cycle as we do, as fertilized eggs, or zygotes. Many biological treatments downplay the rest of the richly varied and crucially

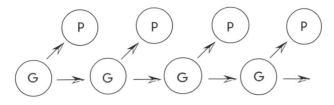

FIGURE 1. *A schematic representation of the Central Dogma of development. G is the transmitted genotype and P is the dead-end phenotype, or organism. Adapted from "Weismann and Modern Biology," by J. Maynard Smith, 1989, in P. Harvey and L. Partridge (Eds.),* Oxford Surveys in Evolutionary Biology *(pp. 1–12), Oxford: Oxford University Press.*

structured contents of the cellular envelope enclosing the chromosomes, so that only the genes are considered to be "passed on." [9] It follows from such logic that the next generation must be made by these genes. This reasoning is often tied to the supposed vanquishing of the Lamarckian heresy that "acquired" characters can be "inherited," but it is just these concepts that are called into question in this book.

The image of the reproductive bottleneck has functioned as a kind of conceptual bottleneck as well. James Griesemer and William Wimsatt (1989; see also Griesemer in press; Sterelny and Griffiths 1999; Winther in press) have discussed the history of the now standard and enormously influential schematic of "Weismannian" genetic transmission, with its linear sequence of genotypes representing the successive generations. In figure 1, bundles of DNA, symbolized by *G*s, are connected by straight lines, forming a continuous lineage. A genotype produces a dead-end phenotype, represented by a branch angling off the main trunk and terminating in a *P*.[10] This diagram is meant to capture contemporary treatments of transmission, not the complexities discussed by the above-mentioned scholars, who point out that Weismann's writings do not support such simple representations. For earlier theorists, *G* would have meant *germ*, and even now *P* could be read as *protein* if one were referring strictly to Crick's (1957) Central Dogma of "information flow" from DNA to proteins (see chapter 3). The point here is that such representations fit quite neatly with the notion of transmission through a narrow aperture.

This phenotype develops and dies, passing only its *G* to its offspring. The significance of the developing organism is thus graphically minimized by showing it as a literal offshoot. At the same time, organisms become mere consequences of the genes' causal powers. The needle's-eye understanding of reproduction and heredity explains how we can have an evolutionary biology that gives such short shrift to the developmental dynamics of the life cycle. Anything other than a gene has as difficult a time getting through reproduction's strait gate as that poor camel in the parable, straining to squeeze through the eye of a needle.[11] And yet, as I show in the chapters that follow, there is only a bottleneck if one accepts a particular narrow definition of heredity. The existence of all those single-celled stages, furthermore, is possible only because they are preceded by, and surrounded by, an enormous functional and structural complexity.

THE DIVIDED *I*

The first two eyes, the blind one turned to development by natural selection and the narrow needle's eye of heredity, set the themes of part 1, "Looking at Development and Evolution." My conviction that evolution's eyes need to be opened, or rather that they actually take in more than most theories do, helps to motivate the arguments in part 2, "Looking at Ourselves." There, two more meanings of *evolution's eye* come into focus. Although present in earlier chapters, these additional meanings are rendered more salient by a shift in orthography. Chapter 6, on contingency, is a "hinge" piece insofar as it joins the use of parity of reasoning in the developmental systems approach to the principle of symmetry in science studies, bringing to the fore the epistemological issues that are especially important in part 2.

My first *I* is the person constituted by a double duality: the familiar nature-nurture dichotomy and the body-mind dualism with which it is closely associated. In this well-worn but still dominant framework, the development of persons must be achieved by somehow joining the natural to the acquired, the biological to the cultural, or, if you will, the body to the mind. I examine aspects of these dualities in part 1, but in part 2 I pay more attention to their consequences for our everyday lives. The widespread and rising interest in evolution has been accompanied by a predictable resurgence of arguments about human nature and basic biological truths. Often, these arguments are also about personal agency and responsibility. Partly natured, partly nurtured, the *divided I* accounts for much of the political charge borne by apparently technical problems in evolutionary theory or genetics. As documented in chapters 7 and 9, discussions of reproductive interests dissolve into arguments over legal culpability, and population geneticists' statistical wrangles become polemics on affirmative action or educational policy.[12] The question of biological bases for homosexuality, discussed in chapter 10, continues to be hotly disputed in the academic and popular literatures, precisely because of its presumed legal and moral implications (Bem 1996; Fausto-Sterling in press; LeVay 1993, 1996). The religious right insists that sexual orientation is a choice not to be explained by biology and points to the testimony of "reformed" homosexuals. In the familiar dynamics of such exchanges, this further increases the pressure on the other side to claim that sexuality is given, not chosen. In chapter 7 I mention the move in some feminist thought from denying fundamental difference to celebrating it. It seems probable that such realignments will increase in the area of sexual orientation (itself a live and still divisive issue in feminism) as well as with respect to human difference and disability in general—as already seems to be the case for certain kinds of deafness and autism.

There are good reasons and not-so-good ones for the persistence of these debates. There is no doubt that important questions can be asked about human possibilities, both collective and individual, about the nature of will, identity, and morality. The questions are difficult to frame meaningfully, much less to "solve." All the more reason to avoid unnecessary difficulties when we can. Several chapters in part 2, particularly chapters 9 and 10, touch on issues of identity and action. These concepts have complicated and interlaced histories. Part of my aim is to address the complications that are built into the distinction between biology and culture, and to see whether the genuine substantive problems can be brought into clearer focus. Faced with a demand to say whether men and women are really different, then, I am more likely to ask, "What do you mean by *really*?" than to provide a definitive answer. Though this can be frustrating if all (!) you want is the right answer, there is something to be gained by pausing, even (perhaps especially) in the heat of the fray, to reflect on what the question actually is, whether there is only one, and what a definitive answer could look like. The more important the issue, the more urgent the need to frame it well, and I believe that the perspective presented here provides a better basis than the traditional ones for individuating precise questions and possible — or sometimes impossible — ways of seeking their answers.

THE KNOWING *I*

The previous reading of part 2's title was the unintegrated *I* of the dichotomies between nature and nurture, body and mind. Knowing, thinking, perceiving, and remembering are all activities of living beings in particular settings, and the other theme of "Looking at Ourselves" is the epistemological aspect of organism-environment interdependence in developmental systems. My interest in the relations between scientists' questions and the phenomena they find, then, while evident in the first part, is more prominent in the second. These issues are taken up in chapters 8 and 11. In the latter I also address the question of what is taken to fall inside or outside legitimate theory. The first-person *I* is also the *I* of the knowing subject, then, who may be, but is not necessarily, a scientist; the emphasis here is on the role of point of view in organizing knowledge.

Besides the familiar binaries referred to above as the *double duality* of nature and nurture, body and mind, are many others that pervasively organize our looking, whether the object of regard is evolution, development, ourselves, or the world in general. There are insides and outsides, autonomy and dependence, necessity and contingency, male and female. Yet, one needn't be a triops to see beyond them.[13] There are excellent grounds for challenging dualisms in biology and its sister sciences. Tim Ingold (1996) has noted that anthropologists rarely think to question the distinction between biology and culture, but the same could be said of the individual-society opposition in anthropology, sociology, and psychology. Much of social constructionism, in fact, not only fails to question these divisions but is actually grounded in them (chapters 7 and 10). Indeed, a poignant aspect of some of this work is that it often begins with a desire to defend the reality and significance of the social against what seems a hypertrophied biology (or individual psychology), but then ends by ratifying some of the very beliefs and practices that feed that biology's unbalanced growth.[14]

By the same token, to inquire whether knowledge is made more by the knower or by the object, even if one's reply places greater stress on the former, may be to rely on the same notions of reified "information" we constantly encounter in biology. The connection is not only analogical; the two discourses overlap because nature-nurture arguments brim not only with genetic information, but also with information from or about the world. The divided organism referred to above, in fact, is frequently thought to be made from these two kinds of information. Textbook discussions of instinct characteristically begin with the rationalist and empiricist philosophers and go on to disputes about how much knowledge is biologically given and how much must be gained through experience. I suspect that many ideas of sensory information (and, while we are at it, of mental representations) are not only as problematic as their genetic counterparts in discussions of development; they are, finally, instances of that same usage.

If information must be understood relationally, as a difference that makes a difference (or, if you will, the information theorist's reducer of uncertainty), then the same constructivist interactionism that integrates insides and outsides in DST's accounts of development should serve for epistemological investigations as well. This kind of construction does not depend on a distinction, say, between the constructed and the pre-programmed, or, for cognition, between cognitive construction and accurate representation.[15] Just as a developmental interaction changes the organism in some way, at some scale (and is ultimately defined by such a change), so experience changes us, for the short term or the long. An informal memory check suggests that once *phenotype* came to refer to the organism, it was restricted pretty much to the body (often just the appearance of the body). More recently, behavior has been included. Can knowledge, thought, indeed, memory itself, be excluded in any principled way?

The Book

I could go on prospecting the interpretive possibilities of my title, but I leave each reader to explore those (perhaps diminishing) pleasures alone.

The present collection is meant to mitigate the inconveniences of the scattered publishing that comes from addressing very diverse audiences. The chapters have been published before, in slightly different form; within each of the two parts they are in roughly chronological order. They have been revised but not updated, though in a few cases more recent work is mentioned in an unnumbered note. I have also changed the term *environmentalism,* which was used in the earlier papers, to *environmental determinism* because today's "environmentalist" tends to be concerned with ecological degradation rather than with environmental shaping of organisms.

Readers interested in the ways ideas and applications take shape over time will see some progressive elaboration and extension of the developmental systems approach. It should also be evident, though, that the precise form of the conceptual problems varies with field and subfield. Part of my point is that what appears to be a unitary problem, say, of "innateness" or "biological base," is in fact an unruly mix of problems that must be articulated and addressed in their own terms.

In order to preserve narrative flow, and because not everyone will read the pieces in sequence, I did not attempt to eliminate all overlap among chapters. It will please me if readers do as I have done: consider these ideas repeatedly, seeing how they play themselves out in different fields, against different backgrounds, and with respect to different problems. In this way their potential reach becomes manifest, but only if they are not mistaken for certain truisms they may appear to resemble (things are connected, nature and nurture can't be *completely* separated, the environment is important, too . . .). At the same time, I will be gratified if, at least some of the time, the reader finds genuine simplicity, the startling kind that comes with a sudden reframing or reseeing. I like it when it happens to me.

Part I

Looking at Development

and Evolution

1 Transmission and Construction:

Levels and the Problem of Heredity

The conventional view of evolution involves two mistaken ideas. One, the idea that traits are "transmitted" in heredity, rests on notions of genetic programming that are ultimately quite preformationist. A second idea, what I call "developmental dualism," holds that there are two kinds of developmental processes, one controlled primarily from the inside and another more open to external forces; it both supports and is supported by the notion of trait transmission. The theory of evolution seems, in fact, to require a distinction between features that develop "under the aegis of the genes" (Konner 1982:157) and those that are shaped by the environment—hence the apparent need for dual developmental processes.

The concepts of trait transmission and developmental duality are linked by a way of thinking about the role of genes and environment in ontogeny that ensures that we will continue to find ways to carve up the living world into innate and acquired portions, no matter how vociferously we declare the distinction to be obsolete. This in turn ensures not only a degree of conceptual incoherence in our science, but also continued difficulty in synthesizing our knowledge of development with our understanding of evolution. Finally, our lack of clarity on these issues, because of their deep involvement in the old and tangled nature-nurture complex, encourages further confusion when scholars make pronouncements on the role of evolution in shaping our fundamental nature (Konner 1982; Midgley 1980; E. O. Wilson 1978). I will argue that evolution only *seems* to require these two ideas. In their place I offer an alternative way of looking at development and the succession of life cycles we call evolution. This alternative way requires no distinction between genetically transmitted traits and traits that develop in some other manner; the distinction becomes unnecessary when we relinquish the conviction that traits can be *transmitted* at all. Nor does it require us to divide psychological characteristics into those that are "biologically based" and those that are due to culture or "phenotypic plasticity" (as in Fishbein 1976:40). What it does require is a conception of development as construction, not as printout of a preexisting code.

Even though the distinction between the innate and the acquired has been under attack for decades (Anastasi 1958; P. Bateson 1983; Beach 1955; Hinde 1968; Klopfer 1969; Lehrman 1953, 1970; Oyama 1979, 1982; Schneirla 1966), and even though it is routinely dismissed and ridiculed in the scientific literature (Alland 1973:14–15; Konner 1982:80–89), it continues to appear in new guises. The very people who pronounce it obsolete manage, in the next breath, to distinguish between a character that is a "genetic property" and one that is only "an environmentally produced analogue" (Alland 1973:18, on phenocopies, discussed in chapter 2), or between "genetically determined fixed action patterns" and patterns in which "innate factors" play but a minor role (Konner 1982: 20–22, 186). Vocabulary and styles of description shift, but the conviction remains that some developmental courses are more controlled by the genes than others.

The reasons for this are many; some have to do with our philosophical traditions, and some with our attachment to certain analytical techniques. Intellectual inertia and the politics and sociology of academic life also help perpetuate dichotomous ways of describing development. In seeking to make our ideas intelligible to students and colleagues, we express them in familiar terms, thus reinforcing old conceptual structures. (Even when we seek to present new structures, we are often *heard* in terms of old ones.) And in justifying our efforts to others, we may point to our work's relevance for accepted paradigms and problems, thus further legitimating traditional modes of thought. Since dichotomies are unpopular these days, having been largely replaced by degrees of biological constraint, open and closed programs (Lorenz 1977; Mayr 1976a), and the like, it is important to realize that such fuzzing of distinctions does not alter the conceptual framework (see chapter 3). It allows people to dodge charges of simplistic determinism without having to change their ways of thinking.

Another potent factor hindering our attempts to transcend the opposition (or even interaction) between nature and nurture is the resurgence of interest in evolutionary questions, among biologists and others. This could have been, and to a limited extent has been, an impetus to serious conceptual reformulation. All too often, though, it has led to the resurrection of largely unreconstructed notions of nature, for reasons both scientific and nonscientific. Our theories are not isolated from what transpires outside the academy, and we must confront the interplay between scholars' ways of thinking and the concerns of the larger society.

Part of the attention attracted by evolutionary studies is prompted by the explosion of theory and research in the field itself. Beginning in the 1970s, we have witnessed several decades of impressive discoveries in molecular biology. Another aspect of the rush to adopt an "evolutionary perspective" is probably a pendulum swing back from the environmental determinism that held sway in the social sciences for many years. In addition, however, there appears to be a more general wish, not limited to academics, for a substantive "biological" bedrock.[1] There seems, finally, to be a hope that an immutable natural "core" will stabilize an increasingly relativistic and uncertain world. Hence the attractiveness of the "ultimate" explanations offered by evolutionary theorists.

These reasons, however, are neither independent nor sufficient to explain the burgeoning interest in evolutionary issues. For now, I want only to point out that they tend to rest on, and ultimately to preserve, largely unanalyzed conceptions of "biological bases," not the least of which is the assumption that "biology" gives us a set of (largely) universal, unlearned, unchangeable, and inevitable traits which were formed by natural selection and which define our fundamental nature. (Context-dependent, facultative variation introduces a wrinkle into this formulation, but it is not a wrinkle that challenges most people's ideas about developmental dualism; see my comment on quadrifurcation below.) It is just this complex of ideas about biology and evolution that needs to be untangled. Although that task requires much more space than is available here, the notion of trait transmission via genetically directed development is certainly an important part of the complex.

Natural Selection and Genetically Programmed Traits

Natural selection, we are told, cannot work on acquired characters, for these are not passed on to offspring; only inherited traits are transmitted. The appearance and differential proliferation of novel genes or gene combinations, coding for new aspects of the phenotype, is seen to be the essence of evolution: Evolutionary processes are said to produce genetic programs for development if one is speaking of the individual organism, or changes in gene pools if the focus is on the population level. Thus, no matter how much one protests that all development requires both an adequate genome and an adequate environment and that all forms are jointly determined by genotype and developmental context, and no matter how often one points out that genetically determined variance is not the same as genetically guided development, the form of certain traits continues somehow to be placed in the genes. A paper by Scarr and McCartney (1983) provides a striking example of this. Writing in the service of an allegedly evolutionary perspective, the authors dismiss the opposition between genes and environment and emphasize the inability of analyses of variance to explain the combination of genotype and environment in development. They are quite happy, however, to speak of genetic programs controlling maturation, to contrast genetic and environmental transmission of characters, and, most emphatically, to speak of the genotype as driving development (p. 428). (They do all this, ironically, on the basis of analyses of variance.)

What has emerged from the confrontation between evolutionary and developmental concerns is a kind of credo that informs our thinking even when it remains implicit. Sydney Brenner (cited in Lewin 1984:1327), who some years ago embarked on the ultimate molecular analysis of a developmental program, offers an explicit statement: "The total explanation of all organisms resides within them, and you feel there has to be a grammar in it somewhere. Ultimately, the organism must be explicable in terms of its genes, simply because evolution has come about through alterations in DNA." In exploring the relationship between embryology and evolution, R. A. Raff and T. C. Kaufman (1983:234) similarly take as their "basic tenet" the assumption that the "genes control ontogeny." This is the case, they say, because evolution consists of genetic changes.

Having accepted the definition of evolution as evolution of genes, these developmentalists have no choice but to conceptualize development of species characters as explained by, caused by, and guided by those genes.

Brenner's case is especially instructive. He set out to describe in complete detail the development of *Caenorhabditis elegans,* a remarkable worm that seems to fit the metaphor of a preset computer program mechanically producing an inevitable result. This worm is quite unusual in the precision of its ontogeny; it always has 959 cells, and the history of each of those cells can be mapped with great confidence. In spite of the prodigious knowledge Brenner and his colleagues have acquired about the ontogeny of this creature, however, or perhaps *because* of it, he now disavows the notion of a program, even warning against loose metaphorical use of the term.

My own inclination is to applaud this repudiation of his original aim. Brenner looked development in the face and reported what he saw: A multitude of events influence each other, set the stage for each other, and run off in improbable sequences, but there was no "genetic program." Cell lineages are unpredictable and complex; by the time one has given a "rule" for such a lineage, one has virtually described the sequence, step by step. There seems to be no logic. The paper by Lewin from which I take this material is entitled "Why Is Development So Illogical?" We cannot, unfortunately, explore the interesting question of what would have constituted a program—presumably a "logical" description—for Brenner's group. After all, we can describe many biological processes in computer language, and we can produce marvelous simulations. But such simulations must provide for much, much more than DNA sequences and their products. More to the point, a program amounts to a description, and we can write these descriptions for all sorts of things, including "acquired" characters. These latter programs may include information on DNA sequences as well. How, then, can the concept of a program help us distinguish between the innate and the acquired?

In any case, Brenner has not lost faith in the power of reductive analysis. In an intriguing (and, to this observer, frustrating) coda, he asserts that he is now convinced that the unit of development is the cell. Because one can presumably understand how the "genes get hold of the cell," this retreat up one level evidently allows him to continue to see development as explicable in terms of genes, even though he pushed this premise to its limit and found it lacking. Having given up genetic programs, he now speaks of internal representations and descriptions. In doing so he is like many other workers who have been faced with the contradictions and inadequacies of traditional notions of genetic forms and have tried to resolve them, not by seriously altering their concepts, but by making the forms in the genome more abstract: not noses in the genes, but *instructions* for noses, or *potential* for noses, or *symbolic descriptions* of them. This solves nothing. At most, it communicates a certain discomfort with the preformationism implicit in traditional formulations, but it does not alleviate the problem; it just makes the wording vaguer.

Developmental Dualism and the Proliferation of Levels

Since so many scientists accept the idea that the body is constructed by genetically guided processes, it is not surprising that they look for parallels to the development of body structure when they turn to psychological and social levels. Some of our mental, even communal, life, they reason, is phylogenetically derived, just as our bodies are, and is therefore attributable to the same kinds of genetically guided development. But insofar as cruder forms of nature-nurture distinctions imply an invariant biological base with environmentally wrought variations, they are incapable of supporting this complication. A consequence is that motivation, thought, perception, and behavior (and eventually culture) are divided into pseudolevels. And so scientists fret about how to identify the inherited aspects of the various human temperaments (Plomin and Rowe 1978; by "inherited," by the way, they do not simply mean "showing heritable variation in a particular population"), how to distinguish biological from cultural universals (Harris 1983), and so on. Both variable and universal features, that is, are somehow divided into those that are (mostly) formed from within and those that are (mostly) sculpted from without. The dichotomous has become quadrifurcate.

But what could it mean to say that behavior of any kind is "phylogenetically derived"? We are told that the behavior is "programmed" or in some other way "in the genes." If we are unhappy with this description, we are hastily assured that no trait, morphological or behavioral, is actually *in* the genes, but rather that information or rules are encoded in the

DNA. Our imaginary interlocutor (imaginary but, I suspect, easily recognized) also assures us that evolved behavior may well be the result of "interaction" between nature and nurture, a mixture of biology and learning, and that *nothing* is due entirely to the genes *or* the environment. And while body structure is "directly" encoded, furthermore, gene-behavior links are "indirect." Any uneasiness we may have about genetic determinism is supposedly laid to rest as our friend assures us that usually the genes do not implacably command, but rather predispose.

If we are uncommonly mistrustful or skeptical, however, we are not mollified. Our doubts are fed by our friend's statements that such evolutionarily shaped behavior constitutes the very core of our being (if we are the species in question); that it is normal, if not desirable; and that it is altered or suppressed only at considerable psychic and social cost (Lumsden and Wilson 1981:358–360; Midgley 1980:75–76, 186). Believing that a particular kind of social behavior has evolved, the friend implies, does not necessarily entail its endorsement, only the sober admission that it is in our nature and is therefore a force to be contended with, to be accepted or ameliorated, even to be vigorously combated within the perhaps narrow limits allowed by the genes, but surely not to be underestimated (Konner 1982:206).

The use of terms like *genetically transmitted* for such features, however, is profoundly misleading. It is just what leads people to proliferate levels within levels as they search for more acceptable ways to express their belief that some aspects of mind, like most aspects of body, are somehow in the chromosomes and can thus be passed on with them. For differential reproduction to alter a gene pool, however, all that is needed is reliable genotype-phenotype correlations; and these, in turn, require not genetic "programs" for development but a reliable succession of organism-environment complexes—of developmental systems that repeatedly reconstitute themselves.

Trait Transmission and Trait Construction

Viktor Hamburger (1980) asserts that developmental studies have not been fully included in the grand neo-Darwinian synthesis. He gives several reasons, one being evolutionists' preoccupation with trait transmission across generations, a preoccupation that is at odds with developmentalists' interest in the elaboration of traits during the course of each life cycle. Developmentalists, he also says, tend to dislike the preformationist implications of transmission genetics. Raff and Kaufman (1983:20–22) describe the estrangement that developed between embryologists and geneticists early in the twentieth century and, like Hamburger, allude to conceptual differences that have hindered the synthesis of the two traditions. The dilemma is how to have an evolutionary perspective without being trapped with some untenable notion of genetically directed development and, by implication, an equally untenable notion of environmentally directed development to complement it. The reason circumlocutions fail to extricate workers from this dilemma is that they are just that, circumlocutions: indirect ways of saying that at least some of ontogeny is explicable in terms of macromolecular plans. No matter how much their own observations contradict this assumption, these scientists are committed to it because they do not question the definition of evolution as evolution of genes.

But the conception of evolution as change in gene pools or genotypes is an exceedingly abstract one—and, more serious, an incomplete one. The role of gene frequencies in reflecting some changes in developmental trajectories is undisputed. The importance of genes in development and as sources of phenotypic variation is also undisputed. What is misguided is the attempt to place phenotypic form, or any kind of "representation" of that form, in those genes. It amounts to an attempt to explain a dynamic multilevel phenomenon by a lower-level entity. (For some discussion of levels in evolution, see Sober 1984.) It is not only reductive, it is incomplete even at the molecular level because it ignores the rest of the interactive complex necessary for gene function: precursors, organelles, and other aspects of cell structure and chemistry. Given that many believe the primary problem of developmental biology is the differential activation of genes at different times and in different tissues, *and given that it is the rest of the interactive complex that must in large part account for differential activation,* this partial account is hard to justify.

To focus on gene pools or modal genotypes as the essence of evolution is to lose sight of the organism, of the successive life cycles whose distribution and reproductive fortunes are reflected in those gene frequencies.

As a gene pool changes, however, so do many of the levels of environment needed to create and sustain those life cycles, and they must be included in any account of ontogeny and phylogeny. They are implicated in even the most basic and reliable of species forms. The environment, that is, must be seen not only as a source of phenotypic variance, but as fundamental to species character, and not only as passive support, but as equal partner with the genes in giving rise to living beings. Vital patterns are the result of interactive systems at many levels.

What is transmitted between generations is not traits, or blueprints or symbolic representations of traits, but developmental *means* (or *resources,* or *interactants*). These means include genes, the cellular machinery necessary for their functioning, and the larger developmental context, which may include a maternal reproductive system, parental care, or other interaction with conspecifics, as well as relations with other aspects of the animate and inanimate worlds. This context, which is actually a system of partially nested contexts, changes with time, partly as a result of the developmental processes themselves. Differential gene transcription in diverse tissues is a result of this emerging system, as are interactions at the organ and organism levels. Developmental means are transmitted in the sense of being made available during reproduction and ontogeny. Often they are the very products of ontogeny, but they are no less crucial to further development for not having been present at conception.

Bodies and minds are constructed, not transmitted. Even what is generally described as culturally transmitted is not simply transferred in a lump from one being to another. Organisms, including people, often influence the development of other organisms; these influences, however, are interpreted and used by their recipient in ways that are complex, ill understood, and very much a function of the organism's developmental state and surrounding conditions. Developmental state, in turn, is the synthesis of past developmental events. We all know the frustration of seeing others construe our words and deeds in ways that seem mistaken, even perverse. What have we transmitted? That which was intended? That which was understood? Haven't we rather entered into interactions whose outcomes are only partially a function of our contributions?

The Construction of Levels

I admit that the term *construction* is problematic in some ways. For one, it often implies an acting subject. While this is perfectly appropriate when we are speaking of knowing subjects construing their worlds and their own actions, it is inappropriate at other levels. We find, for example, descriptions of development or behavior in which genes "foresee" contingencies, "direct" events, "recognize" stimuli, "select" or "program" outcomes. This is construction seen as revelation (Monod 1971:87), as fulfillment of a plan, and it is precisely the construction of ontogenetic "construction" that gets us into so much trouble. My meaning is quite different. Construction, in my view, does not require a subject or subject surrogate, which is often what the gene amounts to in accounts of development (Tobach's 1972 *cryptanthroparion;* Weiss's 1969 *anthropomorphic* principle, or my own 1982 *homunculoid gene*), but it does require multilevel systems constructing themselves, bringing about the conditions for their own further change. I see interactive construction, then, not as the fulfillment of a plan, and not necessarily as the activity of a knowing subject, but as a developmental phenomenon that is not explicable by only one set of its constituents or by only one of its levels.

Developmental Systems

Developmental systems evolve, generating one life cycle after another. One aspect of their evolution is the shifting constitution of the gene pool. But the full phenomenon is a succession of organisms, variable in some ways and constant in others, in their changing environments. In order to put these organisms back into an increasingly gene-centered conception of evolution and development, we sometimes need to step back from the gene level, however fascinating its processes, and reconstitute the rest of the systems that repeatedly make the living world. This is finally what students of development must *do,* however they may *describe* what they do. Whether or not one speaks of genes "getting hold of" cells, that is, one must finally describe not only intracellular processes but also relations among cells, as well as the ways these relations influence and are influenced by higher-level processes, including organism-environment interactions.

The benefits of seeing evolution as a succession of developmental systems are several. First, it restores the organism to us at a time when some biologists seem to be intent on analyzing it out of existence (Dawkins 1982). Second, it allows us to investigate and appreciate the role of the genes in development without turning them into wise little homunculi. Third, it affords us a way out of the multiple versions of the nature-nurture impasse while still allowing us to join our developmental concerns with our evolutionary ones. It does this not by giving us yet another way to make the distinction between genetic and environmental traits (as opposed to genetic and environmental sources of variation), but by showing how *all* features, species-typical or variable, morphological or behavioral, are produced by a complex of interactants, some exceedingly constant, and some variable. This frees us from the bogey of dual developmental processes and from the implicit preformationism and reductionism of all notions of preexisting plans in the genome and restores to us the living world in all its layered complexity without diminishing our ability to turn our analytical skills on any part of that world as we attempt to understand it and to make our way through it.

2 What Does the Phenocopy Copy?

Originals and Fakes in Biology

This chapter examines a concept unfamiliar to most psychologists. The concept of the phenocopy comes from the study of genetics and developmental biology. In psychology, which provides the context for the following discussion, it is occasionally encountered in works on behavior genetics or psychobiology. These are, of course, areas we tend to regard as "biological," and it is precisely the way we construe the biological that concerns me here. The problem is not wholly that of one discipline's ignorance about the data and theory of another; some of the habits of thought I will discuss are widespread among biologists as well.

Genotypes and Phenotypes

A peculiar ritual is observed virtually every time genes are mentioned in the psychological literature. It consists of pronouncing the nature-nurture issue dead, "except in the minds of a few unsophisticated individuals" (Alland 1973:14–15). Alexander Alland, an anthropologist, invokes the principle of interaction between genes and environment, and then proceeds to treat the two as alternative sources of form and function rather than as joint determiners, either of which can be a source of variation. In addition, the genetic is seen as basic, natural, necessary, and more or less fixed, while the environmental is secondary, changeable, and contingent. The first term in "nature and nurture" can be read, after all, as "given by nature" or as "fundamental, inherent." This set of distinctions has a history as long as our culture's, descending as it does from ancient notions of essence and appearance, the eternal ideal and the specific manifestation. Witness one old way of defining genotype and phenotype, less frequently seen now but not yet completely superseded: Nash, in a text on the psychobiology of development, explains that the genotype is "the actual genetic material," while the phenotype is "the appearance of the material" (1970:18). This vision fits very well, of course, with the common account of Mendelian genetics, in which recessive traits go undetected in the first generation, only to reappear in the next.[1] Appearances, we learned, can be deceiving.

Phenotypes, in this view, can "match" (Lindzey, Hall, and Thompson 1978:42) or "correspond with" (Nash 1970:29) the genotype; or they may not. By extension, a phenotype can also match a genotype other than its "own." The phenocopy is generally defined as just this sort of phenomenon: "an environmentally produced analogue of some genetic property" (Alland 1973:18). Norman Munn (1965:26) gives a pair of definitions not generally used these days: Similar traits with dissimilar underlying genes are said to be phenotypes, and similar traits based on the same genes are genotypes. Phenocopies could be examples of the former, though Munn seems to be thinking of the appearance of a dominant character in heterozygous and homozygous individuals. He does not give a term for *dissimilar traits* based on the *same genotype;* as we shall see, this, too is basic to phenocopying.

An exploration of two of the ways the concept of the phenocopy has been employed will reveal not only some assumptions underlying its usage, but also the contradiction inherent in definitions like Alland's. Why would an author pronounce the nature-nurture controversy dead in one breath, and in the next characterize some properties as genetic and others (possibly identical, morphologically and physiologically) as environmentally produced? Surely some hint is to be found in the techniques of experimental analysis, which Alland (1973:15) mentions as useful in separating genetic from environmental effects. As valuable as such analysis can be, it frequently leads to viewing not just variation but also characters, developmental sequences, abilities, and behavior as "genetic," a practice that creates conceptual problems while clarifying little.

Two Uses of the Phenocopy by Behavioral Scientists

The use of *phenocopy* to refer to a phenotype developing under novel environmental conditions and resembling the phenotype associated with some mutant gene arose in a tradition of research in which mechanisms of gene action were sometimes revealed by investigating mutants and attempting to produce the same anomalies in the laboratory. The hope was that the information thus gained might lead to techniques for mitigating or even preventing such "undesirable" effects. The fact that such techniques exist, incidentally, gives the lie to the old assumption that these defects are untreatable. Production of a phenocopy, then, can be the *synthesis* step in the classical research sequence, the step that tests the results of previous analysis. Markert and Ursprung (1971:165–168) describe some of the limitations of such research, which stem partly from the multiplicity of agents that can occasionally induce the same result. Boric acid and insulin, for instance, had the same eventual effect in one case. Not all of these phenocopies are intentionally induced: Phocomelia (limbs resembling seal flippers), for example, is associated in humans with an abnormal genotype or with prenatal development in the presence of the drug thalidomide. Many abnormalities, however, are produced in the laboratory. Rumplessness in chickens can be produced by introducing insulin early in development, while the same agent applied at a later stage produces a different phenocopy, a short upper beak (Markert and Ursprung 1971:166).

In each case, one route to the abnormality involves an abnormal genotype and a normal developmental environment, and the other involves a normal genotype and an abnormal environment. I use *environment* here in a traditionally imprecise and global way, to make a distinction. The *effective* environment for a developing organism, however, is very much a function of its genotype, and to some extent is produced, chosen, and organized by the organism. Topoff (1974:41), in fact, points out the error of speaking of different organisms actually having the "same" environment.

The designation of one of a pair of similar phenotypes as genetic and the other as environmental, then, follows the convention of identifying a phenomenon with the variation that results in its appearance. This is *determination* in the narrow, source-of-variance sense, and it is very different from determination in the broad, formative sense, as is usually intended when "genetically determined" is used interchangeably with "genetically programmed" or "genetically coded." It is the confusion of these two senses of *determination* that lies at the heart of many conceptual difficulties in the behavioral sciences.

Phenocopy also refers to certain alternative phenotypes with the same genotype. The water crowfoot plant has aquatic and air leaf forms; flower color in some plants varies with environmental conditions; the color of certain bacteria depends on the medium on which they are grown (Alland 1973:18–19). And Kurt Stern (1973:84), a geneticist, speaks of a suntan as a phenocopy of dark skin that does not require exposure to the sun. These alternative forms appear normally in nature, so in such cases we do not have an artificial copy of a natural anomaly. What, then, does the phenocopy copy?

In Alland's last example, one strain of bacteria grown on a certain medium is white while a related strain is yellow on that same medium. The white one turns yellow if moved to a second medium. The phenocopy seems, then, to be the yellow bacterium on the second medium, since its color is "environmental," while the yellow of the one on the first medium is "genetic." This "genetic" yellowness is presumably genetic in the sense of being normal for this strain. But the phenocopying bacterium's yellowness could *also* be genetic in this sense, as could its whiteness. The difference between the two colors, furthermore, is environmentally determined in the source-of-variance sense because only a switch of medium is required to produce it. The white form can also be thought of as a phenocopy if there happens to be some *other* white strain that will serve as a "model." Both phenocopying forms, finally, could be models for other phenocopying strains, even though models are "genetic" and copies are "environmental."

It seems that the traditional claim that phenocopies are environmentally produced imitations of genetic traits is based on assumptions of uncertain precision and validity.

What Is Fundamental about Phenocopying?

If there is any substance to the notion of the phenocopy, it cannot depend on an unhelpful distinction between genetic and environmental characters, nor on anything as accidental as the presence of a model. Conrad Waddington (1975:77) makes this clear with respect to the first use of phenocopy described above, and it is equally true of the second. The fact of resemblance seems not to be absolutely basic, in fact, because the term is not used for other forms of phenotypic similarities such as those found among related species, or in cases of mimicry, in which an edible species, for instance, looks like an evil-tasting or poisonous one. What does appear to be conceptually fundamental (at least for my purposes) is the fact of multiple developmental pathways, which is what Waddington was getting at.

The idea that a phenocopy is an environmentally produced phenotype that resembles a genetic one is intelligible only if one believes that there are two sources of development, either one of which can act as an efficient formative agent against the supportive background of the other. Although it is true that the old-style nature-nurture distinction is virtually defunct in scholarly writing, it is still common, as I noted earlier, to find perfunctory but obligatory statements that both genes and environment are necessary to development, followed by blithe designation of certain traits as genetically organized. Developmental processes are then classified as maturational (unfolding automatically, guided by an internal plan) or acquired (formed from without by specific environmental influences). Although combinations and gradations are permitted ("almost purely innate," "partly environmental"), the basic explanatory dichotomy is retained. Because such thinking is common in genetics and embryology (Weiss 1969) as well as in psychology (Hinde 1968; Lehrman 1970; Oyama 1979; Schneirla 1966), biological writings do not necessarily act as a corrective. Lewontin (1985), who is a consistent exception, points out that the difference between a phenocopy and a model is a difference between polygenic and major gene effects, not one between environmental and genetic action. Stern (1973:84) phrases it somewhat differently when he says that phenocopies are "individuals whose phenotype, under the influence of nongenetic agents, has become like the one normally caused by a specific genotype in the absence of the nongenetic agents."

Although some phenotypes are species-typical and some are not (one sense of *genetic*), some are associated with major genes and some are not (another sense), and some characters show heritable variation in a given population and some do not (yet a third), *all* characters proceed from, and are directed and limited by, a given genotype and are thus totally genetic. Each also depends on, and is limited and directed by, a given environment and is thus totally environmental. This is so whether observed variation is traceable to environmental differences or to genetic ones. Each kind of factor, in fact, is often effective only through the other. Genes influence development by altering the immediate environment for various constituents of the organism, while the environment frequently influences the rate and timing of gene transcription. Although it is customary to speak of genes as actively organizing an inert environment during ontogenesis, Yuwiler (1971:51–52) describes the active part played by enzyme products and substrates in "regulating the direction of metabolic flow." They are not to be seen, he says, as simply passive material used by the genes.

Take a child with phenylketonuria (PKU), a metabolic disorder that usually leads to mental retardation if untreated but that can often be controlled by instituting a special diet early in life. The normality exhibited by this child on a proper diet is not an environmentally produced normality that phenocopies genetic normality; it is the result of a particular combination of unusual genome and unusual (for us) environment. The mental retardation of a PKU child on an unregulated diet is similarly the result of coaction, or constructivist interaction; it is no more or less genetic than it is environmental. Either genotype or diet can be seen as the "cause" or determinant of level of intellectual functioning, depending on what comparison one is making and what is being varied. The point is that every genotype has many possible developmental pathways, many leading to the normal range (or ranges, as in the second use of *phenocopy*) and some of them not, and that often a given phenotype may be reached via many different sets of genotype-environment relationships.[2]

Genetic Potential

Everyone "knows" that *any* phenotype, normal or anomalous, depends on a given set of interactions between a specific set of genes and specific environments. Yet such knowledge evidently does not prevent people from believing that some special executive or formative power resides in genes in the guise of a plan, program, or code. In this view the genetic potential of the organism is defined by the cellular code. Environmental conditions can prevent the realization of full potential but cannot overcome the limits imposed by the genes. This is, in fact, the dominant view in psychological writings.

It is trivially true that a response of which a genome is incapable is impossible. Strictly speaking, however, it is the phenotype that responds, not the genes, unless one is speaking of the molecular level at which transcription occurs, and at this level *respond* is probably not an appropriate term. No organism, at any rate, has absolutely unlimited developmental possibilities, and to that extent its "genetic potential" is limited. Similarly, a given environment cannot support all possible developmental sequences. One could say that an organism's *environmental potential* is limited because its genes can operate only within the limits imposed by their surroundings; it's true, but of course one does not say it.

The richness of genotype-phenotype relations is not even approached by the restricted notion of genetic potential; for any organism we know only a small fraction of all possible interactions. For any pair of genotypes, Erlenmeyer-Kimling (1975:25) writes, "the differences in phenotypic expression (or in the phenotypic values of quantitative factors) . . . may widen, diminish, or even reverse ranks from one environment to another" (see also Horowitz 1969). Although the notion of genetic potential is not as rigid as Lehrman's (1970) "developmental fixity," it is conceptually similar. It allows some variation in outcome but implies a fixed "upper" limit dictated by genomic structure. Yet, as Hirsch (1976:163) insists, we cannot say in advance what the norm of reaction will be;[3] it is meaningless to say that the genes set limits to development, because we can never specify those limits.

It is possible that the persistent tendency to attribute developmental fixity of some sort to "genetic" characters derives partially from a failure to distinguish species or populations from individuals. Species-typical characters, for instance, are said to be rigid and difficult to influence. This is taken to mean both that they are relatively invariant across individuals *and* that, once formed in individuals, they are difficult to alter. Or characters whose differences are heritable[4] in a given population are assumed to be difficult to influence in individuals. See, for example, the IQ and race controversy chronicled in Block and Dworkin (1976). The concept of genetic potential figures large in such arguments, as it does in much of the current sociobiological literature.

Astute writers have been saying for years that the discrimination of alternative influences on a population or sample is to be distinguished from the analysis of interacting influences in ontogenesis. Certainly the results of analyses of variance, which involve an inevitably very limited sampling of possible variables and levels, are not appropriately taken as general statements about functional relationships and developmental processes (Hirsch 1976; Lewontin 1976, 1985). Lewontin (1985) further reminds us that the separation of causes in analysis of variance is illusory because the amount of environmental variance depends on the distribution of genotypes, and vice versa.

Phenocopying as an Example of General Developmental Principles

The reality of development, then, is that it is the result of the constant coaction (a term also used in McClearn and DeFries 1973:311) of genetic and environmental factors; to consider some outcomes more or less genetic than others is to fall into a common conceptual trap equally prepared by philosophical and scientific traditions. The special cases we call phenocopies are special not because the processes that produce them are special or even necessarily artificial (recall that alternative phenotypes in the second usage described above occur naturally); they are singled out because certain resemblances have theoretical or practical importance, or possibly because the differences among the alternatives far exceed those usually observed in nature. Such differences are by no means as rare as one might think, however. Even sex, the "genetic" trait par excellence, responds to environmental cues in some species. *Bonellia* is a marine worm with free-swimming larvae. The larvae that eventually attach themselves to mature females are masculinized by hormones produced by the female and become males. Sex changes, behavioral and reproductive, are observed among the coral reef fish *Labroides dimiatus.* The disappearance of the male from a group is followed by increasingly male behavior, including aggressive displays, from the dominant female. Within four days "she" is able to mate with the remaining females, and fertile sperm are produced in fourteen to eighteen days (Lerner and Libby 1976:141). S. J. Gould (1977:chap. 9) gives many examples of species with alternative reproductive forms that appear in response to changes in such variables as periods of light and dark, food supply, and crowding, as well as amphibian metamorphosis that occurs only under certain environmental conditions. In such multiple developmental possibilities lies a good measure of evolutionary staying power, because they offer additional survival capabilities in the face of changing circumstances.

There is evidence that English-speaking psychologists, largely antibiological for so long, are moving away from that stance. This, I think, is a good thing. If it is to be a valuable development, however, the trip back toward the biological must be more than a mere intellectual pendulum swing, a shift of position along the same old conceptual dimensions (more is written into the genetic program than we thought, etc.).

An example of misunderstanding of the biological is revealed by the alacrity with which certain scientists have adopted the term *heritability,* often to fill the vacuum left when *innate* and *inherited* fell from favor. The connotations of fixedness that used to go with the latter two words have frequently been transferred to *heritability.* The polemic on intelligence and race still echoes, along with suggestions that we might, on the basis of low test scores (taken as indexes of inherent potential), foreclose the educational futures of certain children. Other areas, including sex differences, aptitude testing, and, in a more speculative vein, "human nature" as described by some theorists, are also characterized by assumptions of developmental fixity and continuity. It has even been suggested that *heritability* should replace *instinct* (Whelan 1971).[5]

In deploring such thinking I am not making a claim for infinite technological efficacy or unlimited "malleability"; the point is not that everything can be changed or improved, but rather that responsiveness to various kinds of influences is a function of phenotypic state and of the particular way the influence is brought to bear. We need to be modest about the scope of our knowledge and to avoid affirming null hypotheses about developmental possibilities on the basis of "genetic components."

If we accept the idea of multiple developmental pathways that is implicit in the notion of the phenocopy, we will be in a position to discard traditional unilinear conceptions of development and the expectation of relatively simple continuity that often accompanies them. Students of development are beginning to move away from reflex assumptions of long-term predictability (Clarke 1978), although the belief in fixed potential, whether rooted in the genes or in early experience, dies hard.

What is to be gained from biology, then, is not a growing inventory of genetically coded traits, as some would have us believe, but a set of principles, methods, and data to be used in exploring the questions of psychological development, stability, and change. To use a biological perspective is thus not to discuss body type, for example, by conceding that a "genotypic ectomorph" might *look* like a mesomorph and then going on to wonder what the behavioral phenotype of such a "mesomorph-mimicking ectomorph" might be, as Nash (1970:88) does. It is rather to ask whether the types form consistent classes with reliable psychological correlates, and if they do, to investigate the conditions for their development, the relationship among the types at various developmental stages, the conditions under which an individual might change his or her classification, and so on. Similarly, asking whether aggression or altruism or hope or incest taboos are encoded in the genes (and, of course, if one is "thinking biologically," answering that indeed they are) is less productive than critically examining the concepts and finding out how and when and by whom the various phenomena are exhibited, to what situational and ontogenetic variables they are responsive, how to measure them, and so on.

So What *Does* the Phenocopy Copy?

If all that lies ahead is biological reductionism, then I do not greet the new day with enthusiasm. If, on the other hand, we can extricate ourselves from the intellectual muddle created by reifying potentials and taking misleading metaphors literally, we can get down to the business of finding out how things work and how they come about.

S. J. Gould (1977:395), in his analysis of the relationship between ontogeny and phylogeny, suggests that much of the saga of evolution will eventually be explicable not in terms of "the extreme atomism of 'bean-bag' genetics," which posits an "independent efficient cause" for each feature and explains it by citing adaptive utility, but rather by discovering the mechanisms of developmental regulation. The differences between humans and other primates may, he says, be less a matter of a genotype augmented by genes "for" human qualities than one of retarded development tempo resulting in neoteny (p. 405).

If we are interested in information and instructions, we need to look not only at the genes but also at the various states of organism and the ways one state is transmuted into the next. Potential is probably more usefully conceived of as a property (if it can be thought of as a property at all; see above comments) of the phenotype, not the genotype. It is the phenotype that can be altered or not, induced to develop in a certain direction or not; its potential changes as each interaction with the environment alters its sensitivities (Oyama 1979). A disorder may or may not be associated with a major gene, Thiessen (1972:91) asserts, but therapeutic intervention is related to the phenotype, "not to the point of gene impact." Rodgers (1970:213) makes a similar point.

In explaining why "information" directing ontogenesis does not reside entirely within the developing cells, Klopfer (1973:27) urges us to think of the cell as "an information *generating* device, not an information *containing* device. Its ability to generate information, . . . to produce certain compounds, depends on the immediate environment being organized and predictable." One need hardly add that different information will be generated under different circumstances. Klopfer goes on to say that the stability of development depends on feedback systems which, being self-correcting, allow much more effective regulation than would be possible if "instructions" were contained in the cell, "written once and then read off without change." It is further true that the "setting" of a biological feedback system may be altered by interchanges with the environment; potential is thereby changed, and subsequent exchanges may be regulated in a very different way.

Far from being environmentally determinist, this view acknowledges all contributions to development, both typical and atypical, while agreeing that in any given situation, variation may be traceable to any of a number of factors. One must remember that what makes a difference is very much a function of what is varied and what is held constant, and at what levels.

What does the phenocopy copy? Another phenotype, of course, but when similar phenotypes result from the development of organisms with quite different genotypes, who is to say which is the original and which the imitation? Is one more or less an expression of genetic potential than the other? Each "copies" the other. And when a given genotype gives rise to dissimilar phenotypes, how do we decide which is the one that is genetically coded? It's time to stop shuttling between the two determinisms. There is not only more than one way to skin a phenotypic cat, there is sometimes more than one way to *become* one.

3 Ontogeny and the Central Dogma:

Do We Need the Concept of Genetic Programming

in Order to Have an Evolutionary Perspective?

The title of this chapter implies that it is possible to have an evolutionary perspective without the concept of genetic programming (and without any of its surrogates, which proliferate wildly in the psychological and biological literature). This is indeed the case, but before considering how it can be accomplished, and why it is important, we would do well to ask why evolution and programming have been assumed to be inextricably joined.

My discussion begins with some remarks on the need to integrate evolutionary and developmental studies, two areas that have been estranged from each other for some time. The rift is hardly surprising: Evolutionary theory has been associated with a view of development as centrally controlled and predetermined. This is a view rejected by many who are deeply appreciative of the interactive systems that generate living forms, but it fits an old tradition of preformationist thinking, and is thus difficult to give up. I then look at the standard definition of evolution as changes in gene frequencies, a definition that seems to require genetic control of ontogeny. Finally, I reformulate the concepts of inheritance and ontogeny, and argue that the idea of the developmental system allows us to have an evolutionary perspective without being saddled with an untenable doctrine of one-way flow of developmental "information" from the nucleus to the phenotype.

Knowledge and the Shadow Box

SEEING

Howard Gruber and Isabel Sehl (1984) studied the ways people cooperate to construct knowledge that is not available to either of them alone. Pairs of people were presented with a box in which shadows of an object could be seen. Because each person looked at the object from a different angle, each saw a different shape. One might see a circle, for example, whereas the other might find a triangle. Together, the partners had to construct an object that could project both of those shapes. Clearly, some ways of working were more effective than others: Domination was not particularly helpful in moving beyond a one-sided view, and mindless compromise in the absence of real constructive work (the object is partly round and partly triangular) was inadequate as well.

Like the partners at Gruber and Sehl's task, developmentalists and evolutionists have often had difficulty communicating with each other. Sometimes the two groups have had divergent interests, and the language in which they have communicated has prevented them from realizing it. It is possible that their difficulties are also related to the presuppositions they bring to that shadow box we call science, and thus to what they are able to construct there. Viktor Hamburger (1980), as noted in chapter 1, has described the exclusion of embryology from the neo-Darwinian synthesis. The lack of accord between developmentalists and evolutionists on the evolutionary role of embryological development has had a partial analogue in studies of behavior. Witness the exchanges in the 1950s, 1960s, and 1970s between American comparative psychologists and European ethologists (Lehrman 1953, 1970; Lorenz 1965). To some extent these groups talked past each other because they were interested in different matters. But I think they also had genuinely different conceptions of development.

FOUR POINTS OF VIEW

In 1963, the ethologist Niko Tinbergen suggested that when someone asks why an animal does something, there are four possible biological interpretations of the question. Many students of animal behavior have been guided by his explication of the "four *whys*": (1) the evolutionary history of the behavior, (2) its survival value, (3) the mechanisms by which it occurs, and (4) its development. Tinbergen thought that failure to distinguish among these questions had caused considerable confusion. It still does. Developmental psychologists have not traditionally thought much about the first two, and when they have, they have not always kept their *whys* straight. Nor have they necessarily been helped by their colleagues in biology. Too often, the price of admission to the biological brotherhood has been a view of development marked by reliance on assumptions that are decidedly uninformed by systems thinking, despite a vocabulary liberally sprinkled with systems terms (e.g., Fishbein 1976; Scarr and McCartney 1983). The notions of genetic programs and instinct, for instance, draw some of their authority from their associations with evolutionary thought. Programs and instinct carry all sorts of other implications about mode of development and kinds of mechanism, however—autonomy, internality, spontaneity, naturalness, resistance to perturbation—neatly tying together the four *whys* that Tinbergen distinguished. (He did not claim that they were unrelated, just different. He even discussed some ways of relating them [1963]; see also P. P. G. Bateson 1985.)

Imagine for a moment two people at a shadow box. This time they have been given different instructions: One partner is to discern the object's shape, and the other must determine whether or not it is moving. Imagine also that they must use an ambiguous vocabulary. *Rooving*, for instance, means both "round" and "moving." One person reports that the object is round, and the other concludes that it is mobile. I am suggesting, in this crude way, that their plight resembles that of scientists who must work with ambiguous terms such as *inherited, genetic, biological,* and *maturational.*

The ambiguity of these terms permits the blurring of Tinbergen's *whys.* If one worker discovers that a bit of behavior is present in phylogenetic relatives, for instance, the other may conclude that it will be developmentally stable. If one claims that a pattern is adaptive, the other may deduce that it develops independently of experience. Evidence for one *why* masquerades as evidence for another. If their aims and terms were clarified, the researchers might conclude that they were pursuing separate projects. But the goal is not separatism; it is rather the recognition of differences that is the prerequisite to fruitful collaboration. A great deal of conceptual work is required as well. Integration, after all, is not the same as conflation. Furthermore, because these notions of "biological bases" seldom segregate neatly into developmentalists' and evolutionists' heads, coherence within fields suffers even if there is no interest in interdisciplinary work.

Evolution and the Central Dogma

THE ARGUMENT

The usual way of construing the relationship between ontogeny and phylogeny involves several interrelated ideas. First, evolution is defined by changes in gene frequencies. The genes are thought to produce phenotypes by supplying information, programs, or instructions for the body and for at least some aspects of the mind. Some genes produce better phenotypes than others and are differentially passed on. Although in this view the environment is necessary for proper development, its effects on the phenotype are evolutionarily insignificant because only inherited traits are transmitted. Causal power and information are carried in the DNA, and living things are created by an *outward flow* of causality and form from the nucleus.

This conception of the ontogeny-phylogeny relationship seems to call for two kinds of development: one for inherited traits and one for everything else. This is true even though statements are routinely made about the impossibility of attributing traits *completely* to one or the other. Despite their reassuring ecumenical ring, such statements either retain the dichotomy or turn it into a continuum. Emblematic of a trendy but failed "interactionism," they are responses to a multitude of developmental observations that call traditional formulations into question; their shortcomings are examined later. (For critiques of this kind of well-intentioned but conceptually misguided interactionism, see Lewontin, Rose, and Kamin 1984; Oyama 1982, 2000; Tobach and Greenberg 1984.) Although they scornfully dismiss "extreme views" that attribute behavior *entirely* to the genes or *entirely* to the environment, the devotees of this popular interactionism mistake compromise and relabeling for conceptual resolution.

Many have expressed their unhappiness with the contradictions and faulty inferences that accompany these accounts of the ontogeny-phylogeny relationship, and have tried to formulate a unified conception of ontogeny (P. Bateson 1983; Gottlieb 1976; Johnston 1987; Klopfer 1969; Lehrman 1970; Schneirla 1966; Tobach 1972). Not surprisingly, they have often had difficulty communicating effectively with colleagues who hold the dominant developmentally dualistic view. Indeed, it can be argued that the nature-nurture dichotomy will continue to dominate our theories and research as long as we continue to speak of traits, programs, or encoded potential as being *transmitted* (see chapter 1).

THE GOAL

We need to alter our conceptions of ontogeny and phylogeny before we can bridge Hamburger's (1980) "nucleocytoplasmic gap." We do not need more conciliatory declarations that nature and nurture are both important, but rather a radical reformulation of both. All too often, as we shall see below, people confuse the genotype with the phenotype. Those who manage to avoid that oft-sprung trap may still become obsessed with genes because they have been encouraged in innumerable ways to think of them as prefiguring or making the phenotype. I propose the following reconceptualizations, in which genes and environments are parts of a developmental system that produces *phenotypic* natures:

1. *Nature* is not transmitted but constructed. An organism's nature—the characteristics that define it at a given time—is not genotypic (a genetic program or plan causing development) but *phenotypic* (a product of development). Because phenotypes change, natures are not static but transient, and because each genotype has a norm of reaction, it may give rise to multiple natures.

2. *Nurture* (developmental interactions at all levels) is as crucial to typical characters as to atypical ones, as formative of universal characters as of variable ones, as basic to stable characters as to labile ones.

3. Nature and nurture are therefore not alternative sources of form and causal power. Rather, nature is the *product* of the *processes* that are the developmental interactions we call nurture. At the same time, that phenotypic nature is a developmental resource for subsequent interactions. An

organism's nature is simply its form and function. Because nature is phenotypic, it depends on developmental context as profoundly and intimately as it does on the genome. To identify nature with that genome, then, is to miss the full developmental story in much the same way that preformationist explanations have always done.

4. Evolution is thus the derivational history of developmental systems.

CONDUITS AND MESSAGES

George Lakoff and Mark Johnson (1980), a linguist and a philosopher, present an "experientialist" alternative to what they call "objectivist" and "subjectivist" theories of knowledge. Properties of objects are produced in interaction rather than residing in the objects or, alternatively, being completely arbitrary and subjective, and metaphor plays a central role. Lakoff and Johnson's attempt to provide a third way, a synthesis that transcends a traditional antithesis, has striking parallels with my attempt to use constuctivist interaction to move beyond nativism and environmentalism.[1]

Lakoff and Johnson (1980:10–12, citing Reddy) describe the "conduit metaphor" for language: Ideas or meanings are objects that can be placed in the containers we call *words* and sent along a conduit (communication) to a hearer. The meanings then reside in the sentences and are independent of speaker or context. The objectivist theory of communication is based on this conduit metaphor: Fixed meanings are sent via linguistic expressions (Lakoff and Johnson 1980:206). In a similar way, I suggest, natural selection is thought to place knowledge about the environment (or instructions for building organisms that are adapted to the environment) into the genes, which then serve as vehicles by which these biological meanings are transmitted from one generation to the next. The context independence of meanings in the conduit metaphor is consistent with the connotations of autonomy and necessity that accompany the ideas of instinct and genetically driven development. (On context sensitivity in systems theory, see Valsiner 1987.) The details of the communication metaphor tend not to be well worked out in biological discourse (Johnston 1987; Oyama 2000), but I think the parallels are robust.

Perhaps biologists' propensity to speak of molecular letters, words, and sentences, of genetic codes and grammars, is to some extent the result of notions of life as language. The genes become the repository of true nature; molecular meanings are contained in a phenotypic vessel, which is sometimes treated with so little regard as to render it virtually transparent. A few paragraphs ago I mentioned confusion between genotype and phenotype. This may have stuck some as too bizarre to be credited. Yet consider some psychologists' practice of effectively bypassing the phenotype: Plomin (1986:110, 129) speaks of people responding to children's "genetic propensities" and "genetic differences" rather than to the children themselves. Similarly, Scarr and McCartney (1983:433) refer to experiences that "the genotype would find compatible." The latter authors use the term *developmental system,* but their insistence that genes and environment play quite different roles in this "system" reveals the distance between their understanding of this term and my own: In their scheme, of course, the genes play the determining role. (Despite the importance of both, that is, some causes are more equal than others. For contrast, see Fogel and Thelen 1987; Valsiner 1987, on systems dynamics.)

The Central Dogma: Hypothesis and Metaphor

Genes appear to link evolution and development in two ways. First, they are the material link that promises to make sense of heredity. The adoption of Francis Crick's (1957) Central Dogma of the one-way flow of information as the ruling metaphor for development forged a second, conceptual link. The dogma states that information goes from genes to proteins, never from proteins back to genes; as metaphor, it takes many forms (programs, blueprints, instructions; see Newman 1988), but they always involve the emanation of basic developmental causation from the DNA. An outward flow of information and power achieves the translation of the genetic message in ontogeny.

A subtle transition is thus made from "messages" about molecules to messages about bodies and minds—quite a different thing, whether we realize it or not. The shift is from *gene* transmission to *trait* transmission. Focus on the gene as the prime mover of ontogeny leads to all sorts of assumptions about genetic control of development as the defining characteristic of certain traits; this in turn leads to the need for *another* kind of process to explain everything else.

The more reductionist one is, the harder it is to appreciate the gap between the molecular and the organismic levels. The kind of reductionism I am speaking of here is not the provisional single-mindedness that allows detailed investigation of a mechanism. It is rather the desire to interpret the whole world in terms of that mechanism, or at least in terms of the level at which it was studied; it is the failure to shift levels or point of view, whether from inability or from some conviction that to do so would be soft-headed. The genetic program holds a fatal attraction for such minds.

Crick's Central Dogma has come to have the quality of an unquestionable truth. We forget that it is a hypothesis about specific molecular interactions that is open to empirical support or refutation. But what could challenge the "plain truth" that development is controlled by a genetic program? What would count as evidence for or against such control? One biologist told me that "the whole of molecular biology" demonstrated the reality of the genetic program — hardly the language of normal scientific inference. Genetic programs have so dominated mainstream thinking that people have rarely been called on to defend them. Programs have functioned as an "enabling concept," important not only in research but in the legitimation of a kind of reductionist explanation (Yoxen 1981:105). Richard Dawkins (1986:111) speaks of DNA as programs, instructions, and algorithms, and explicitly denies that these terms are metaphors. Of airborne seeds he says, "it is the plain truth [that it is raining instructions]. It couldn't be any plainer if it were raining floppy discs." Organisms become "natural-technical objects structured by logics of domination," "biotic components in a technological communications system," "command-control systems" (Haraway 1981–1982:247, 259, 271).

When *system* is simply shorthand for "machine governed by a program," it usually signals a preoccupation with static centralized control rather than the sort of distributed, dynamic, contingent control under consideration here. (For an analysis of the popular use of *systems* and its links to a preoccupation with control, see Rosenthal 1984:chap. 13.) We thus have an uneasy association of systems terms with a style of explanation that is, as previously noted, uninformed by the kind of systems theory that offers the most to developmental studies.

In addition to asking what kind of evidence is relevant to program explanations, one could ask another question: Does the notion of the program *add* anything of value to the understanding gained by analysis of developmental processes? I submit that it does not, and worse, that it usually imports extraneous and misleading implications. The Central Dogma as metaphor helps make intuitive sense of observations, fitting them into a particular worldview. Metaphors are not mere embellishments to thought; they are fundamental to knowing itself. I would hardly argue, therefore, that we should give them up. But not all metaphors are equally useful, and one that encourages us to see development as the fulfillment of a plan or the transmission of a message may not always direct our attention in desirable ways. In addition, the program metaphor supports some of the more troublesome aspects of the nature-nurture opposition. Despite the comfortable fit between the idea of evolution as evolution of genes and the idea of genetically programmed development, there are profound problems with this formulation.

Two Strategies

Two general strategies can be used in arguing for a genetic program. One is to say that some features develop by means of a genetic program, whereas others do not. This is the conventional dualistic formulation. The other is to say that *all* development is in some way controlled by the program. I call this "genetic imperialism" because it appears to be an attempt to include in the genes' purview both the "innate" and the "acquired": to subdue the environment and the organism's history in it. It, too, is developmentally dualistic in attributing different causal roles to the genes and the environment.

CONVENTIONAL DUALISM

The problem with attributing some parts of the phenotype to the genes and some to the environment is that developmental processes and products are simply not partitionable in this way. Nor are the various criteria for making the distinction consistent. Although the concerns motivating any particular nature-nurture distinction may be interesting and important, casting the question in terms of the two constrasting factors of "nature" and "nurture" immediately joins it to a multitude of other questions that have radically different empirical bases. For example, presence at birth, appearance without obvious learning, longitudinal stability, reliable timing, and susceptibility to perturbation are quite distinct developmental issues, and adaptiveness, phylogenetic relationships, and patterns of distribution in a population are not developmental questions at all. As I argue in chapter 7, lumping all the issues together as manifestations of the same thing, namely genetic nature, guarantees that the conceptual chaos at the shadow box will continue, veiled by a common vocabulary.

Thus one finds blithe cross inference from populations to individuals and back again, from development to evolution and vice versa, from adaptiveness to mechanism, from phylogenetic similarity to necessity and naturalness, and so on ad infinitum, all because terms like *biologically based, genetically encoded,* and *inherited* give the illusion of movement within a coherent theoretical system. Jacobson et al. (1983:436) declare that if adults' tendency to address infants in high-pitched voices is species-typical and adaptive, then it is "biologically programmed in the adult speaker" and is thus neither learned nor responsive to feedback from the infant. Conversely, if experience *can* influence a behavior, Connor, Schackman, and Serbin (1978) conclude, then it is *not* "biological." Frodi and Lamb (1978) reason that certain human sex differences are not biological because the behavior changes over the life span. But the multiple properties attributed to "biology" are not inherently linked to each other and therefore cannot be inferred from each other. Beards, breasts, and reproductive behavior are not present at birth; some learning is species typical; adaptive characters are not necessarily unchangeable; and universal characters by definition show no heritable variation.[2] Characters shared with phylogenetic relatives do not necessarily appear early and are not always difficult to change. I could go on.

Assuming we have an adequate definition of learning, we can certainly ask whether a given behavior is learned or not, or whether it is influenced by prior learning or not. To answer such questions, however, we need to look at the *development* of the behavior, not ask whether it is difficult to change or wonder whether it is universal. Traits that are reliably found in a species can be distinguished from those that are not on the basis of the reliability of the various aspects of the developmental systems that produce them. Species-typical influences may be typical because they are passed on in the germ cell, because they are part of a larger reproductive system, because they are created or sought by the organism itself, because they are supplied by conspecifics or other organisms, or because they are otherwise stable aspects of the niche. These associations must be investigated if we are to understand the differences between uniform and variable traits; our understanding is not improved by the circular tactic of explaining observations or conjectures by varying the amounts of genetic programming necessary for their occurrence. The prediction of the future presence of a trait, which is often the issue in the nature-nurture debate about humans, is not properly accomplished by identifying the trait with the genes (by computing heritability coefficients, by detecting it in baboons or in hunter-gatherers, by declaring it adaptive, and so on) but rather by understanding the developmental system well enough to allow us to say whether the entire system, or an equivalent one, will inevitably be present in the future. Careless inference not only hinders investigation of the question at hand, it also prevents a satisfactory integration of development with evolution. Although slips like these are sometimes discovered and corrected by vigilant scholars, what is needed is correction of the conceptual system that generates them in the first place. Later, I will point out the missed opportunity that such confusion represents, and argue for a radical reconceptualization.

GENETIC IMPERIALISM

Conventional dualism fails, then, because it rests on an incoherent mix of ideas; there is no consistent way of distinguishing features that are programmed from features that are not. Replacing the dichotomy with a continuum (some traits are more genetically programmed than others) does not solve the problem; the same inconsistencies are found in such conciliatory-sounding formulations as in strictly dichotomous ones. The other way of construing the genetic program is to declare that the genes determine the range of possibilities: They set the limits on development. As in conventional dualism, the mechanisms whereby the genes supposedly do this are obscure, but this does not deter some from claiming that norms of reaction are genetically determined (Freedman 1979:150; Mayr 1961). According to Scarr (1981), the range of reaction is the "expression of the genotype in the phenotype" (p. 16), and the "genotype has only those degrees of freedom that are inherent in its genes" (p. 17). Although she claims that it is incorrect to say that "heredity sets the limits on development" (p. 17), that is exactly what is entailed by the notion of genetic degrees of freedom. This is all quite ironic because the array of phenotypes that could be associated with a given genotype is just the array in which all differences are *environmentally determined*. The environment, after all, is seen as "selecting" the particular outcome. The norm of reaction is therefore a nice demonstration of the joint determination of the phenotype. Every organism incorporates "information" from genes and environment in a complex that cannot be partitioned as variance is partitioned. But such mundane truths do not seem to be the point here, which is rather a kind of metaphysical urge to contain ontogenetic variety within genetic boundaries.

The problem with this imperialistic version of developmental dualism is that it is vacuous; a genotype has just those developmental possibilities that it has (though who is to say what they are). Used this way, the program no longer has empirical content. It is more like a symbol of ultimate faith. Or it may be only a fancy way of saying that potential is finite. In fact, one variant of this idea is "programmed potential": Mayr (1961:1502) claims that "the range of possible variation is itself included in the specifications of the code" of the genetic program. But because the range of possible phenotypes is defined by the set of genotype-environment *pairings,* what is the point of attributing that range to just *one* member of the pair? And why insist that the range be fixed at fertilization? Potential must be a *developmental* concept if it is to be useful. It cannot be treated as a fixed quantity somehow inscribed in the genome (Horowitz 1969; Lewontin 1984). As many have noted, it is just this idea of fixity that has led people to draw conclusions about things such as intellectual potential from heritability figures.

A given genome may certainly have several developmental possibilities. But those possibilities vary with the developmental state of the organism and the context. Traditional notions of maturation, readiness, and embryonic competence turn on the realization that possibilities must emerge in ontogenesis. A bee larva has at one moment the potential to become a queen or a worker but a short time later the worker-to-be may no longer aspire to royalty. Its genes are the same, but its effective potential has changed. Similarly, as I noted in chapter 2, a dominant female cleaner fish ordinarily looks forward to an entire life as a female. Should the male in her group die, however, she becomes a fully functional male within weeks (Lerner and Libby 1976). Potential, then, does not always diminish over time. A phenotype *develops,* a fact that the concept of genetically programmed potential purports to explain but actually ignores.

One can, of course, attribute to the genome a higher-level potential for all these potentials, but this is only to affirm that all possible developmental outcomes are possible. It is also worth noting that such a move does not help to distinguish adaptive outcomes from nonadaptive ones, normal ones from pathological or idiosyncratic ones. The claim that the genes circumscribe potential reminds me of a ploy used by the powerful when they realize that power must be shared, if only minimally: Delimit the scope of choice, then let the other party choose within fixed, nonnegotiable boundaries. "It's time for bed; which pajamas do you want to wear?" [3]

Deep Problems, Superficial Solutions

The difficulties with the conceptions of development discussed above have not gone unnoticed. Changes have tended to be cosmetic, however, for a number of reasons.

SHARED BELIEFS

Evolution, first of all, is so firmly identified with the genes—gene pools, selfish genes, genetic information—that any attempt to call attention to other aspects of evolution or to question conventional definitions of inheritance is immediately seen as some sort of Lamarckian attack on scientific biology. (Sapp 1987 discusses geneticists' success in imposing genecentric definitions of evolution and heredity, showing how these scientists used the image of a genetic control center simultaneously to describe cellular activity and to suggest their own powerful position in science.)

In addition, so much thought and research are rooted in the nature-nurture tradition that it is difficult to think differently. "The biological" is seen to be more real, more basic, more normal, more recalcitrant to change than "the psychological" or "the cultural." We have already seen that the various criteria for designation as biological or inherited do not form a coherent set; we have also seen how easily their incompatibilities are glossed over when they are referred to with the same terms. Scholars and laypeople alike continue to distinguish necessary inner essence from contingent outer appearance. Consider the oppositions that have been so important in the behavioral sciences: instinct versus learning, maturation versus experience, inborn personality traits versus acquired ones, to name but a few. Kenneth Kaye's (1982:4) list of the "great issues" that psychologists have attempted to resolve by studying human infants ranges from the pedagogical to the theological, and they are all cast in traditional nature-nurture terms. If one such distinction is questioned, minor local adjustments may ensue, but the complex interweavings of these ideas in our thought and practice make more serious change unlikely. Frequently, the *versus* is changed to *and* or *interacts with* and the problem is considered solved. (Many examples of this are found in Magnusson and Allen 1983, in which the authors refer to combinations of innate and environmental factors, the interaction of biology and learning, and so on, all in the service of an "interactional perspective.") Although the cooperation implied by these phrases is more pleasant than the oppositional tone of earlier formulations, the dichotomy remains. Hence the need to distinguish my views of development from this kind of conventional interactionism. Given the broad ramification of those habits of thought, it is no wonder that small changes to vocabulary or theory are so easily assimilated to dichotomous views.

Two kinds of phenomena might have posed a serious challenge to the logic of programming explanations: species-typical learning and adaptive variation. But these "bugs" in the concept of the genetic program were integrated into traditional thought with minimal discomfort.

"PROGRAMMED" VARIATION

Complex behavior that is difficult to account for by learning (e.g., instinct) has traditionally been "explained" by the genes. So has reliable, closely orchestrated development (e.g., maturation). Although the concept of instinct has often been questioned, that of internally driven maturation has largely escaped scrutiny, probably because nativists and empiricists alike shared basic beliefs about physical development (Oyama 1982). Even the most dedicated behaviorist requires a body and a reliable set of operants and reflexes to begin a conditioning story. Yet, researchers eventually realized that some learning is species-typical, and indeed is crucial to many "instincts" (avian imprinting is the classical case, but see also Hailman 1969).[4] This threatened two traditional renderings: of instinct as unlearned and of learning as arbitrarily variable. The resolution proposed by some theorists, however, involved not a reconsideration of the notion of genetic programming, but an *increase* in the scope of the program. Not only are bodies and unlearned behavior in the genes, now some learning is in there, too.

These theorists' attempts to replace the innate-acquired distinction with closed and open programs or with inherited ranges of possible forms (Lorenz 1965, 1977; Mayr 1961), or innate, genetically determined epigenetic rules (Lumsden and Wilson 1981) merely blur the distinction between traditionally conceived nature and nurture when they should be questioning the very basis of that distinction. Fishbein's (1976) description of "canalized" learning as genetically preprogrammed development is typical of this approach, but many others have made similar attempts to reconcile species-typical learning with the ideal of genetic control (Freedman 1979; J. L. Gould 1982; Shatz 1985; and see Johnston's 1988 critique of innate templates in avian song learning). Perhaps these efforts are unsurprising in light of the fact that *species-typical,* and more often, *species-specific,* became euphemisms for *instinctive* and *innate* in the discourse of workers who tried to take critiques of instinct into account but did not realize that conceptual tightening can take more than an adjustment of the lexicon. (An even more recent example of such hedging is the ubiquitous term *constraint,* discussed in chapters 4 and 5.) Such maneuvers unfortunately tend to be presented as the leading edge of developmental theory, where an "evolutionary perspective" too often means making more and more refined nature-nurture distinctions while attributing more and more to the formative, directive powers of the genes. These theorists claim to be eliminating the nature-nurture dichotomy, but in reality they are simply renaming it and shifting phenomena from one side of it to the other.

Parallel with the inclusion of some learning in the concept of instinct, some investigators (Lorenz 1977; Mayr 1976a) saw that any idea of species-typical development would have to include the possibility of branching pathways to accommodate certain kinds of adaptive variation (alternative morphologies or behavior patterns). External events, that is, often intrude into the supposedly autonomous maturational sequence to move the organism onto one or another path.

So we see that learning can be necessary for the development of behavior usually defined by the absence of learning (instinct), and divergent pathways can be crucial for development usually considered unilinear (maturation). This shows the impossibility of consistently categorizing developmental phenomena as innate or learned. Putting them on a continuum defined by varying amounts of genetic control does not solve the problem; it multiplies it. The hazards of switching from one definition of innateness to another were pointed out above in the discussion of Tinbergen's *whys;* in some of the works cited previously we see the consequences of confusing innate as species-typical, innate as predetermined, innate as conferring survival advantage, innate as unlearned, innate as having an evolutionary history, and innate as independent of the environment. These failures of consistency could have challenged the conceptual framework of developmental dualism. Instead, however, the offending phenomenon in each case was simply assimilated to the old system, giving us genetically programmed learning and genetically programmed developmental branching. The apparent adaptiveness of these violations of ontogenetic autonomy, as well as their selectivity and presumed evolutionary histories, compelled workers to find a way to attribute them to the genes, even though this necessitated finessing the Central Dogma of development (that "the biological" is created exclusively by the *outward* flow of genetic information). These efforts, however well-meaning, are finally just superficial responses to profound conceptual problems. That such phenomena can be treated nondualistically is evidenced by Caro and Bateson's (1986) analysis of alternative tactics.

LOCAL AND GLOBAL CHANGE

We need to rethink such category-defying phenomena as well as the reasons for, and meanings of, the various kinds of inquiry. Even minimal rethinking can help, but piecemeal progress is risky. The unexamined assumptions we sweep under the rug will trip us up as soon as we turn around.

Eliminating the kinds of unjustified cross inference just described would be an example of low-level improvement. Although Lamb et al. (1985) characterize their research on parenting in human males as biological, for instance, they emphasize variation with context, not fixity. They rightly deny that evolution must bring invariance and immutability. One wishes for more, however. They do not abandon the notion of "hardwired predispositions," and term them *physiological* (p. 886). What happens when sex differences, for instance, are discovered by methods like hormonal assays or evoked potentials? Are they then "hardwired" because physiological? What would count as a *non*physiological "predisposition"? Unanalyzed terms like these are the lumps lurking under their carpet, inviting us all to stumble.

Local improvement is of local utility. Change on a broader front involves reworking whole networks of concepts and whole patterns of reasoning, not just refining a term here and there. This means going beyond formulations like those of Lamb et al., who comment that social conditions can either "override" biological predispositions or "reinforce" them (1985:888), and who present biology and the environment, as so many do nowadays, as "complementary" (p. 886). Broader change would also involve clarifying the scope of research. Lamb et al. (1985) claim to be following Tinbergen, but by treating behavior as "decisions" made in some context and based on the goal of maximizing fitness, they treat an evolutionary function as a proximate cause, an error Tinbergen (1963) explicitly warns against. They give no evidence that fitness is actually maximized by the variations in human parenting they review, and do not make clear just what they hope to establish. Having relinquished the definition of biology as fixity, they seem to be uncertain about just what it *does* mean. Evolved behavior may indeed vary with context, but this certainly does not mean that all behavior that varies with context is evolved in the sense of having a traceable history in phylogenetic relatives.

In a more sophisticated account, Kaye (1982) speaks of aspects of the social environment as being inherited by an infant. He notes that much experience is ensured by evolutionary history, and describes early development as being intensely social. He asserts that it is often informative to look at the child as a part of a larger system, in which an adult may play "cognitive" roles that the child is not yet ready to perform. (See also Rogoff and Wertsch 1984, on Vygotsky's concept of "zone of proximal development.") Kaye sees many universal skills as constructed in interaction rather than revealed in maturation. Yet he is quite happy to speak of the abilities the infant brings to early infancy as innate, maturing "according to the designs of the genetic program" (p. 17). He explicitly excludes from psychology the study of maturation, which is "guaranteed by the genotype" (p. 28). (The problem is not that he attributes the wrong things to maturation, but that he attributes maturation and innateness to a genetic program.) He also criticizes those who liken development to a train ride "in which the very process of the journey is determined by its destination . . . because it suggests that the child knows where he is headed" (pp. 15–16). Ironically, the intrinsic genetic program is an explanation-by-destination in which it is not the child who knows where he is headed but his genes.

Kaye (1982:31) rejects some nativist accounts of human development, charging that those who overestimate the functions present at birth "try to explain away some of the mysteries that have led so many psychologists to begin looking at infants in the first place." But the concept of innateness does just that. (Kuo also made this point about the concept of instinct, in 1922.) Kaye's assertion that infants "inherit certain aspects of their social environments as much as they inherit their nervous systems" (p. 8), although a provocative step in the right direction, reveals the problem. In my terms, he has mixed developmental *influences* (what I call *"interactants"* or *"means"*) with developmental *products*. Genes and social environments are inherited interactants, available to be used in constructing a life cycle. Nervous systems and social skills, being phenotypic, must develop.

Kaye associates evolution with innateness and so is forced to circumscribe his "interactionism." (*Innate behavior* is cross-referenced to *evolution* in his index.) It follows (1982:24–25), then, that the "genetically determined behavioral tendencies" of parents are inherited biologically, while the rest is inherited by the mechanisms of cultural evolution. Kaye's attempt to reconcile the role of experience in development with evolution is a commendable one. If he had applied his constructivist thinking to the concepts of innate behavior and maturation as well as to social development, his account would not have required the developmental dualism that now permeates it. This chapter contains many examples of nondualistic research on the developmental role of naturally occurring experiences. It is true, as Kaye says, that developmental psychologists seek a set of "givens" with which to begin their inquiries, but there is a difference between taking a set of abilities as given, or present, *at some age* and attributing them to a particular kind of developmental process.[5]

Development and Evolution

If evolution is construed as change in the constitution and distribution of developmental systems, the study of ontogeny is no longer a poor relative, to be lent evolutionary legitimacy by genetic hook or crook. Rather, it becomes the very heart of evolutionary biology. And because transgenerational stability and change depend on the degrees of reliability of developmental processes and a large array of means for repeated ontogenetic constructions, research on the processes responsible for transgenerational continuity is crucial. West and King (1987) articulate an idea of the "ontogenetic niche" that is very close to my developmental system (see also Johnston 1982, and the "developmental manifold" of Gottlieb 1971). "Ask not what's inside the genes you inherited, but what the genes you inherited are inside of," West and King (1987) advise; the niche that the genes "are inside of" is an indispensable bridge between generations, and research on the details of that bridge makes up the first body of research to be sketched below.

The second broad research strategy to be discussed involves linking development with other *whys* in Tinbergen's list (see Klama 1988). For this project, though, the usual markers of inheritance and innateness will not suffice, for they are part of a tradition of reasoning that has outlived its usefulness. No more attempts to distinguish features formed in phylogeny from those formed in ontogeny. No more searches for genetic plans for morphology or behavior.

Both strategies, looking at developmental links between generations and relating development to the questions of function or evolutionary history, can generate fascinating research. The projects are not mutually exclusive, and both involve a willingness to investigate phenomena that tend to disappear when the language of programming is used. Indeed, one could say that whenever a program is invoked, a developmental question is being ignored, or worse, being given a spurious answer.

LINKS AMONG GENERATIONS

When constructive interaction is seen to be fundamentally important for the formation (not just the support) of *all* features, including "biological" ones, then the role of the environment is not *complementary* to that of biology, but is *constitutive* of it in much the same way the genes are. This allows a more global reorientation to living organization, one that goes far beyond the local improvements just cited. Attention can be focused on the way any influence is (or is not) integrated into a developmental system, rather than on partitioning the organism according to the role "biology" (however construed) is imagined to play in forming it. Later, I describe the successive levels of developmental systems, from the nucleus out. Any research that sheds light on the origin of novelty at any level could potentially help us understand how variant systems come to be. Any research that shows how processes can be faithfully repeated across generations could help us understand how systems persist. Most of a developmental system remains unchanged in the face of evolutionary alteration. The genetic links between ontogeny and phylogeny mentioned earlier are necessary but not sufficient: After all, the genes alone cannot give rise to the next generation. Although many life cycles narrow to a single cell (Bonner 1974), and although theoretical accounts often reduce that cell to naked DNA, the developmental system is much more extensive. *Its ramified complexity and reliability are just what allow such drastic narrowing of the organismic part of the cycle.*

A first step in extending the developmental system beyond the gene is appreciating the inheritance of complex cellular structures and constituents (Sapp's history of research on cytoplasmic inheritance, which includes, but is emphatically not limited to, cytoplasmic genes, is called *Beyond the Gene* 1987). Recognizing the integration of mammalian embryonic development into the maternal physiological system is a second step (Cohen 1979; Hofer 1981b:224, on the mammalian mother as "an external physiological regulatory agent" —external to the infant, but a crucial and very reliable part of the developmental system, regulating, and regulated by, the developing infant). Looking at the dependence of development on the organism's own activity and its interactions with conspecifics takes us even farther out. Biochemical and even social interaction with parents or siblings can begin before birth or hatching (Gandelman, vom Sall, and Reinisch 1977, on the influence of fetal position on development of mice; Gottlieb 1978, on the effect of prehatching experience in ducklings) and can obviously be of great importance later as well (Johnston and Gottlieb 1985; Lickliter and Gottlieb 1985).

Of great utility here is Gottlieb's (1976) concept of bidirectional relationships among gene action, physiology, function, and social influences. The principle is nicely demonstrated in Vandenbergh's (1987) account of the regulation of mouse puberty by other mice. Vandenbergh's paper also serves as a model for relating individual and population levels. A good presentation of many infant-parent relationships is found in Cairns (1979), which shows the multiple ways investigators may fruitfully move among different fields. That book is also full of illustrations of the social embeddedness of development, and thus of the ways that similarity across generations may be maintained or compromised by interactions among organisms. This embeddedness is also evident in research on the transgenerational perpetuation of behavioral sex differences in rodents (Moore 1984) and of food preferences and other behavioral patterns in a wide range of species (Galef 1976). Similarly, Trevarthen (1982:77) shows how human infants' "mental partnership with caretakers" extends their abilities to act. He comments ironically on the idea of the "isolated thinker" and declares that infants "must *share* to know" (p. 81).

These are all examples of developmental research that highlights the *connectedness* of the emerging organism to its surroundings, not its insulation from them. Although such attention to developmental context provides an excellent vehicle for the study of individual variation, it does not limit researchers to such study. Indeed, it can also show the ways in which specieswide patterns of development are maintained by stably recurring contexts, *and how these patterns of development play a role in maintaining that very recurrent stability.*

It seems quite possible that much of this research comes from its authors' appreciation of systemic complexity and their realization that one must move an investigation beyond the boundaries of the organism in order to understand the organism fully. This ability to see links and intimate interchanges with the surround as developmentally fundamental is not part of developmental psychology's conventional focus on individuals, despite frequent references in the literature to "transactions" between children and the environment (or between genotypes and environment, Scarr 1981, or between nature and nurture, Plomin 1986:20).

Often, such work on developmental interactions is seen as environmentally determinist or behaviorist, and thus opposed to "biological" approaches (Furth 1974; J. L. Gould 1982, 1985; Lockard 1971; Lorenz 1965:3–4). These charges obviously rest on developmental dualism. Although it is true that this research shows the importance of environmental structure, as well as highlighting many possibilities for developmental variation, it is in no way anti- or nonbiological. On the contrary, it illuminates the very phenomena that programming language "explains away"; it shows some of the many ways biology *works*. It certainly does not signal a belief in blank slates, though one must be able to lay aside some very well rehearsed scripts in order to see this. Indeed, to move toward a systems view one must realize how bad the image of a slate really is. The conflict over whether it is environmental features or genetic messages that are impressed on the organism reveals the profound similarities between empiricist and nativist views. Both are committed to a notion of development as imposition, not interactive emergence.

The association of biology with necessity and uniformity is indeed mistaken, as the believers in programmed variation realize. If we are to understand how uniformity and variation are constructed in real lives, however, the metaphor of the program, the internal inscription, is no substitute for real investigation. I doubt that the sorts of research mentioned here could have been conceived by people who were still in thrall to dualistic thought. Although they vary in the consistency with which they avoid nature-nurture oppositions, these workers all appreciate the fundamental role of organism-environment interchanges in the most basic developmental processes.

LINKS TO FUNCTION AND EVOLUTION

We turn now from ecological, physiological, and behavioral links between the generations to theoretical links among Tinbergen's four *whys*. If questions about immediate causation and development are clearly distinguished from questions about evolutionary history or survival advantage, it is possible to seek ways to relate them to each other. Ronald Oppenheim (1980), for instance, has shown the functional significance of many ontogenetic phenomena (see also Turkewitz and Kenny 1982). Patrick Bateson (1979, 1984, 1987; Caro and Bateson 1986) has consistently interpreted development functionally, as have most of the workers cited in the preceding few paragraphs.

Knowledge of the natural history of the species in question is clearly useful in making such functional connections. When we are that species, however, special problems arise. Not only is human variability notorious, but the whole notion of a single natural history for our own species is equivocal. Charting a path through contemporary and historical variation in ways of living is simple only if one is willing to ignore a great deal and to make some arbitrary choices. The preoccupation with reliable life cycles has too often been part of a desire to discover a single, transcultural, and ahistorical human nature, a "biological base" that would unify diversity. But Voorzanger (1987a:51) points out that evolutionary history does not provide a conception of human nature or give us moral guidance. Instead, "we have to know ourselves in order to give an evolutionary reconstruction of our behavior."

Seeing natures as developmental products, and thus as phenotypic rather than genotypic, turns us away from the search for transcendent reality and back to the processes and products of development. Much work remains to be done on the proper relationships between data and constructs in these investigations, and it is my conviction that the nature-nurture opposition, long a dominant heuristic in many fields, is more often a hindrance than a help in this endeavor.

Interaction in Ontogeny:
Sources of Variation, Sources of Form

Current notions of genetic information are unable to account for single developmental pathways, much less alternative phenotypes. Under developmental analysis, any ontogenetic course resolves to multiple pathways at the cellular level. In normal embryological differentiation a single genotype is involved in the development of many types of cells and organs; again we have a kind of "norm of reaction." The *variation* in cell types is "environmentally determined" (involving the immediate environments of the genes and cells), and all *outcomes* are jointly determined as developmental processes generate a multitude of effective genotypes and transient environments—that is, DNA sequences and contexts that interact in a spectrum of particular developmental circumstances. The genotype-phenotype mapping is complex, contextually and developmentally contingent, and, to some extent, indeterminate.[6]

Similar problems exist with the notion of information in the environment. A given event carries, or, better, generates, different "information" for different organisms, and for the same organism in different states. The tenderly proprietary smile is at one moment a welcome sign of love, whereas a year later it threatens entrapment. On one day a gaping chick provokes parental feeding, whereas a month later the same gape stimulates a reaction that means, loosely, "Go feed yourself." In each case the "information" conveyed by one organism depends on the context and on the history and state of the organism that is interpreting it.

The only way to use the idea of developmental information effectively is to detach it from the notion that ontogenesis is a conduit for the transmission of messages. Developmental interactants are "informational" not by "carrying" context-independent messages about phenotypes, but by having an impact on ontogenetic processes—by making a difference. Sometimes those differences are perceptible in a naturally occurring array, as they are in a set of clones developing in different environments. In other cases the arrays must be created experimentally; this is the way contributions to normal development are usually investigated. The research of Gottlieb and his colleagues (1976; Johnston and Gottlieb 1985; Lickliter and Gottlieb 1988) on nonobvious influences on development, including self-stimulation, shows how inadequate it is to regard the developmental environment as supportive but not formative of species characteristics. Much earlier in development, electrical currents generated by the embryo seem to be an important form of self-stimulation (Jaffe and Stern 1979).

Information is a difference that makes a difference (G. Bateson 1972: 315). The concept of the developmental system allows us to distinguish between genetic and environmental variation that makes a difference (generates developmental "information") and variation that does not. But the distinction can be made only with reference to the rest of the system, and thus may vary with it. We can speak, then, of genetically or environmentally determined *variation,* but not of genetically or environmentally determined *traits.* The fact that the difference between two cell types is due to extracellular conditions does not make the cells themselves "environmentally determined" any more than a trait that shows heritable variation in a population is "genetically determined." Phenotypic "messages" are constructed in interaction. This is true whether we study species-characteristic development, as Piagetians do, or species-variable ones, as students of temperament or personality do. Constructivist interactionism, that is, should not be associated only with variability or mutability. It is not, as previously noted, a code word for environmental determinism; nor does it signify some overarching preoccupation with "plasticity."

Interactions between chemicals, between tissues, between organisms, and between an organism and the inanimate environment are parts of the developmental system, and the immediate context of the interaction may be as important as the identity of the interactants. For some animals, "context" is not restricted to physical environment. Interpretation of the situation is crucial. Consider the change in effective context when a subject of the television program *Candid Camera* realizes what is afoot. Or think of the dilemma of the psychologist who wonders whether subjects simulating some social process are a good model for what occurs outside the laboratory. The question is really about what situation the subjects are *in,* and inspection of the room, even through a one-way mirror, will not necessarily give the answer.

The vocabulary of interactionism has been widely adopted, but the full implications of a constructivist interactionism have not been accepted nearly as readily as its terminology. Taking interactionism seriously means rejecting the Central Dogma as a metaphor for the control of development, even for development of the body. (Notice that this metaphorical notion of information flow is independent of the question of reverse translation in molecular biology, which is the empirical issue Crick was addressing.) The one-way causation it implies is inconsistent with the reciprocal, multiple causation actually observed in vital processes. Interaction requires a two-way "exchange of information": Genetic and organismic activity are informed by conditions, even as they inform those conditions. This is nothing more than the bidirectionality so commonly invoked by developmental psychologists today. My complaint is not with the concept, but with the fact that often it is not taken seriously enough.

What Is Inherited?

Traditional gene-for-trait language implies a kind of preformationist embryology, and so do many updated, facelifted versions. But does the inheritance of discrete genes entail the inheritance of traits, in what Cohen (1979) calls the "jigsaw" model of development?[7] My answer is obviously that it does not, but the language of selfish genes, of genetic programs and encoded traits, tends to collapse the distinction by treating the gene as a homunculus that makes body parts and mental structures according to a prior plan.

Some have suggested that we handle the growing dissent over these issues by further separating evolutionary questions from developmental ones (Dawkins 1982). It is true that questions about evolutionary adaptiveness are not the same as questions about how a particular adaptive structure is constructed in ontogeny. But the fields have been estranged for too long. The solution is not to keep them apart even more assiduously, but to synthesize them by shifting our focus from the gene as the unit of evolution and the agent of programmed development to the concept of the evolving developmental system. That is, we must widen our concepts of inheritance and ontogeny to include other developmental interactants as well; no organism can develop without them all. Evolution involves change in the system constituents and their relations.

TRANSMISSION

A major step in the opening out of crucial concepts is the reconsideration of the notion of transmission. Accounts of gene-culture coevolution (Boyd and Richerson 1985; Durham 1979; Lumsden and Wilson 1981) use the model of trait transmission for culture as well as for biology, seeking to remedy the shortcomings of purely genetic theories. By adding a second transmission channel, however, they also continue the dualistic tradition that ensures those shortcomings. In addition, they retain, and extend, the population geneticists' habit of taking genes out of organisms and placing them into mathematically manageable "pools," concentrating on the countable while taking for granted the processes that generate and regenerate these countable entities. (See Keller 1987 on the way this tactic serves the ideology of individualism.) Now we have pools of cultural bits as well, and the repeated and varying life cycles of the organisms themselves are treated as virtually epiphenomenal *effects* of the differential propagation of units from these two pools. But as I have insisted, traits are not transmitted, developmental influences are. Our inheritance does include culture, not as a second set of traits transmitted via an extragenetic conduit, but as aspects of the developmental context. Hofer (1981a) points out the inadequacy of the ideas of genetic and cultural evolution in accounting for all sorts of prenatal effects, and Voorzanger (1987b) maintains, as do I, that an evolutionary theory that included a sufficiently rich account of development would have no need of a second transmission system. In fact, the very idea of transmission would be transformed, because an adequate account of development renders the conventional conceptions meaningless.

The transmission metaphor denies development. If it is development that we are interested in, then we should choose a vocabulary that takes it seriously. Other people's ideas, actions, values, habits, and beliefs are part of the rich complex of developmental influences from which lives are constructed. So are the genes, and so, as noted here, is much, much more. Whether or not any given *trait* will be reconstructed in any particular generation is a contingent matter, for it depends on the constitution and functioning of an entire system. Stability of species characteristics is the result of stable developmental systems. This does not depend on absolute reliability of all interactants; some processes are stable despite considerable variation in their constituents, and some outcomes may be stable despite variation in process.

Developmental systems are to some extent hierarchically organized. They can be studied on many levels, and relations among the levels are crucial. (See Salthe 1985 for an attempt to formalize relationships among levels.) Developmental biologists speak of cytoplasmic inheritance, which can involve extragenetic changes in cell structure capable of being propagated in a lineage (Cohen 1979; Sapp 1987). Even though *variations* in a cell can be inherited in this way, *invariant* features of cell structure are passed on, too, just as the genes common to an entire species are passed on. The rest of the normal environment is also quite reliably present (that is what it *means* to be a normal environment), and when it is not present, some other environment is.[8] Biologists also speak of cell state being stably "inherited" (Alberts et al. 1983:835); the key here is not change in genetic material, but transgenerational (here, generations of *cells*) stability of the cell type. There are, in other words, both species-typical aspects of developmental systems and variations in those systems, genetic and otherwise, and an organism inherits the entire complex. The fact that we daily acknowledge the indispensability of both genes and surround to the development of all characteristics and yet continue to attribute some of that development mostly to the genes and some to "other factors" suggests that our theoretical vocabulary has not kept up with our observations. Both evolutionary and developmental studies remain largely genecentric (Goodwin 1984).

What passes from one generation to the next is an entire developmental system. Heredity is not an *explanation* of this process, but a statement of that which must be explained (see my previous discussion of links between generations). The concept of evolving developmental systems gives a unified view of development while integrating it with evolution. Dualism is no longer required; the inherited-acquired distinction, *as long as it is construed as a distinction between kinds of developmental processes or sources of form,* can be eliminated—not modified or turned into a matter of degree, but eliminated.[9]

Ontogenetic means are inherited; phenotypes are constructed. This enlargement of the idea of inheritance seems outrageous to minds trained to identify it with the genes, but I am only making explicit what is routinely taken for granted. No one claims that genes alone are sufficient for development, or denies that environments, organic and inorganic, microscopic and macroscopic, internal and external, change over organismic and generational time. What is missing from most accounts is the synthetic processes of ontogenetic construction.[10] Inheritance is not atomistic but systemic and interactive. It is not limited to genes, or even to germ cells, but also includes developmentally relevant aspects of the surround—and "surround" may be narrowly or broadly defined, depending on the scope of the analysis. Inheritance can be identified with "nature" only if it embraces all contributors to that nature, and nature does not reside in genes or anywhere else until it emerges in the phenotype in-transition. Nature is thus not properly contrasted with nurture in the first place; it is the product of a continual process of nurture.

Having redefined inheritance in this way, we must also redefine ontogeny. This is difficult for behavior scientists, who are used to squabbling over degrees of biological programming of personality or behavior but who tend to take programming of the body for granted. Although there may be doubt, that is, about whether sex roles and aggression are in the genes, there is usually no doubt at all that sex organs, teeth, and claws are. (Recall the incomplete interactionism discussed above under "Deep Problems, Superficial Solutions"). But a unidirectional flow of genetic information doesn't account for a tooth or a claw, red as it may be, any better than it accounts for the most idiosyncratic behavior. The developmental system, on the other hand, accounts for the emerging phenotype in a way the naked genome cannot.

DEVELOPMENTAL SYSTEMS

Some fear that the concept of the developmental system requires them to give up too much; in fact, it only eliminates a troublesome set of assumptions and inferential habits. Many of the issues formerly associated with the false opposition between nature and nurture (or biology and culture) can still be addressed, but this time clearly formulated and properly distinguished from others. Some are largely evaluative (many concepts of normality, for instance) and are answerable not solely by empirical investigation but by moral discourse as well. If this conceptual unpacking is performed, we will be less likely than we have been in the past to make ungrounded predictions, to draw illegitimate conclusions from our data, to posit distinctions where there are none.

The developmental system is a mobile set of interacting influences and entities. It includes all influences on development, at all levels of analysis. Any particular investigation will obviously focus on a limited portion of the system. For an embryologist, the scale of cells and organs defines the investigative field, and higher-level aspects of the system can generally be taken for granted. In some cases, though, as we have seen, it becomes useful to pay attention to other factors as well, for the wider environment may also intrude; witness the effects of radiation or of various chemicals on embryogeny, or other ways in which the experience of a mother may affect her offspring or even grandoffspring (Denenberg and Rosenberg 1967; Hofer 1981b:chap. 10).

Even though it is easy to think of perturbations of the early developmental system (often mediated through the parents), it is essential to realize that the aspects of the environment that do *not* vary are hardly rendered developmentally irrelevant by virtue of their reliability. Gravity usually does not vary, but in its absence bone and muscle may atrophy, possibly because the pituitary produces insufficient growth hormone (Anonymous 1985). The relevance of this finding to the possibilities of life in space is obvious, but we are well advised to let it act as a conceptual reminder as well. Many of the broader ecological factors (topography, atmospheric composition, patterns of vegetation, temperature, and humidity) have changed over evolutionary time, sometimes as a result of their interaction with life forms. The system changes over the life cycle and is reconstituted in successive generations in ways that are similar to, but not necessarily identical with, preceding ones. This is the only way to have inheritance of genetic material (and other interactants) without being stuck with inheritance of traits.

Examples of interactants in developmental systems include the following (additional references on these topics can be found in "Links among Generations," above):

1. The genome, whose parts interact and move about in ways now being described by molecular biologists (Dillon's 1983 book is called *The Inconstant Gene;* see also papers in Milkman 1982).

2. Cell structure, including organelles, some of which have their own distinctive DNA, and seem originally to have been internal symbionts (L. Margulis 1981).

3. Intracellular chemicals, some of which (e.g., messenger RNA from previous generations) may allow considerable developmental progress before the organism's own genes are transcribed at all (Raff and Kaufman 1983).

4. Extracellular environment—mechanical, hormonal, energetic—parts of which, like the extracellular matrix, are created by the cell itself or by other cells.

5. Parental reproductive systems, both physiological and behavioral; prenatal effects are common, and cross-fostering experiments can

show dramatic effects of parental behavior, sometimes to the extent of producing behavior quite atypical of the strain or species (Hofer 1981b:chap. 11).

6. Self-stimulation by the organism itself.

7. Immediate physical environment, including provisions left for young, as when eggs are laid on or in a food source.

8. Conspecifics and members of other species with which important interactions take place. This category includes, but is not limited to, symbiotic relationships.

9. Climate, food sources, other aspects of the external environment that may influence the organism, initially through the parents and later directly.

In many life cycles, a variety of factors can bring about the major branchings of the developmental pathway discussed earlier. Other influences contribute to less dramatic variations, including variations in learning and anything else that helps define the norm of reaction. These factors are also part of the developmental systems of these organisms.

Nails for Shoes, Nails for Battles, or Why We Need the Concept of the Developmental System

An old maxim goes, "For want of a nail the shoe was lost, for want of a shoe the horse was lost, for want of a horse the rider was lost . . ." Loss of a particular rider might even lead to losing an entire battle. To know whether that nail makes the difference between losing and winning the battle, shouldn't we know what kind of battle it is, on what terrain it is being joined, what the command structure is, and who the opponent is? Even if it could be shown that the loss of a battle were traceable to a lost nail, this would not make the nail an adequate causal explanation for the entire complex of events that constituted the battle. Indeed, it is the entire complex that defines the nail's role.

The nail can become a nail "for" losing battles only in a world that is sufficiently stable and integrated that the entire battle—to say nothing of the geopolitical circumstances that led to it—can be re-created with some regularity. A gene is a gene "for" a given phenotypic difference only if other aspects of the ontogenetic complex (not to mention population structure) are fairly stable. Similarly, an environmental feature can act as a developmental trigger only if the system is competent. If these various conditions are reliable, one can take them for granted and predict outcomes with some certainty even without understanding the developmental processes involved. Much nature-nurture questioning can be translated into queries about the constitution and degree of reliability of developmental systems. But a system that is well integrated in some worlds, and that thus tends to appear as a unit in those worlds, may not be similarly unified in others.

Evolution is only partly a matter of changing gene pools. It is also a matter of changing developmental contexts, and one cannot be understood without the other. Niches evolve in several senses. Geological, climatic, and organic features of an area change over time, partially as a result of the resident organisms. Nor are niches definable apart from their organisms (Johnston and Turvey 1980; Lewontin 1982), and as a lineage evolves, so do its relations with its surround. The niche is the *effective,* the developmentally or functionally *significant* environment; an organism may exploit the "same" environment differently at different times.

We return, then, to the struggling, squabbling scientists peering into the shadow box, trying to make sense of their conflicting accounts. If an ambiguous vocabulary and confused concepts are hindering our communication, we must at least clarify our terms. Our eventual goal is the integration of diverse points of view, both within and among people, by cooperative construction. As useful as it often is to use and reuse a heuristic, sometimes it is necessary to break set, restructure the cognitive field, and move on.

We must relinquish the Central Dogma's one-way flow of causality, information, and form as our guiding metaphor for development. The same is true of the programming metaphors we have taken from computer technology. We must also give them up as the principal framework for our research and interpretation. They do not do justice to any careful investigation of a developmental process, whether at the level of macromolecules or of individuals. Although they provide a familiar and comfortable way of interpreting the world, they have outlived whatever usefulness they may once have had.

Because it is so deeply rooted in our thought and practice, the nature-nurture complex, more than other faulty scientific frameworks, has significant political and moral repercussions far beyond the research community. It has multiple sources in our philosophical history. It influences the classification of individuals in ways that profoundly affect their future. It influences our view of what is possible for individuals and for the species as a whole. Thus, it affects the manner in which we mobilize for maintenance or for change—indeed, whether we mobilize at all.

Hamburger (1980) says that embryology was not integrated into the neo-Darwinian synthesis that had apparently unified so much of biology for three principal reasons. First, evolutionists tended to focus on the outward flow of influence from the nucleus, whereas embryologists focused on the cytoplasm as crucial in determining differential gene activation. Second, the preformationist implications of the notion of particulate inheritance made embryologists uneasy. Third, evolutionists emphasized trait transmission across generations and neglected trait elaboration over the life cycle. These factors are still barriers to effective communication between at least some developmentalists and some evolutionists. They raise thorny problems for those who try to be both at once; as we have seen, nucleocentrism is problematic even for those who have no special interest in evolution. Hamburger calls for an interactive view to bridge the nucleocytoplasmic gap into which so much misunderstanding and acrimony have been spilled. I believe this need is met by the constructivist approach described here. This approach offers a way of speaking about complex transgenerational continuity and variability, about stability and change in both species and individuals, while allowing us to acknowledge the intricacy and contingency of the processes observed in ontogeny, a way to think in evolutionary terms without being committed to a developmental dualism in which contingent nurture is pitted against genetically predetermined nature. We do not need the genetic program in order to have an evolutionary perspective.

4 Stasis, Development, and Heredity:

Models of Stability and Change

Although contemporary evolutionary theory is said to rest on a synthesis, it is also based on a number of antitheses. Internal and external forces are opposed in explanations of ontogeny and phylogeny, biology is contrasted with culture, organisms are separated from their environments. Insofar as development is attributed largely to internal forces and evolution to external ones, these too are contrasted. But development, as I said in chapter 3, is the very heart of evolution. To see why this is so, we must examine our assumptions about the nature of ontogenetic and phylogenetic change and adopt a more dynamic, holistic approach to biological processes.

I define evolution as change in the distribution and constitution of developmental (organism-environment) systems. This often involves change in gene frequencies, but focusing exclusively on the gene level excludes from life processes the very richness and activity that commanded attention in the first place. Defining evolution, heredity, and development in terms of genes also commits us to the nature-nurture oppositions we have been examining. This is especially the case when the organisms in question are human beings, and when biology is identified with "propensities" that seem to leave little room for deliberation, choice, and action.

This chapter looks at two models of change and at three other concepts that are closely tied to them. The models have been called "variational" (differential perpetuation of fixed variants in a collection) and "transformational" (predetermined change in an entity or a collection of entities) (Lewontin 1982). The concepts they inform are: (1) natural selection, (2) innateness, and (3) hereditary transmission of traits.

Contemporary thinking about each of these three is dominated by a ruling metaphor. Natural selection is treated as action by an external agent. Innate traits are attributed to internal "programs." And, finally, heredity is modeled on the transmission of wealth and objects in human societies—thus, traits are "passed" from one organism to another. These metaphors distort the phenomena they describe and create a spurious set of connections among them. Because heredity is seen as the transmission of "genetic" traits, or of genes "for" traits, evolution is reduced to changes in gene frequencies, and ontogeny must be largely explicable in terms of those genes. The whole story requires that the genetic products of evolution pass through the bottleneck of a narrowly defined heredity; they can then re-create the living world by ontogenetic expansion. As in algebraic expansion, however, the solution is implicit in the first equation and must simply be *revealed* by routine operations. The idea of genetically created phenotypes then reinforces the idea of natural selection as "operating" on static traits. We thus have a kind of reverberating circuit of ways of thinking about life processes. This circuit must be broken if we want a fresh look at the relations among selection, development, and heredity. My aim in this chapter is to explore the aspects of evolutionary theory that imply passivity and stasis in organisms, and to show that these implications can be denied without misrepresenting biological processes. We can make development central to evolution without being committed to notions of predestined change in either ontogeny or phylogeny.

Two Models of Change

Richard Lewontin (1982, 1983a) has described two models of change that have been treated as alternative explanations for development and evolution. Each combines two ideas that are not necessarily connected. In each case, I suggest, it is possible to accept the first idea and reject the second.

In the *transformational* model of evolution: (1) change in a collection is explained by change in its constituent entities, and (2) change in the entities is predetermined and uniform.

In the *variational* model: (1) only some of the variants in a collection are perpetuated, and (2) the variants themselves are static.

Change is assumed to be internally driven in the former model and externally directed in the latter. Notice, though, that these assumptions stem from the second component in each—*predetermined change* and *static variants,* respectively. When these are dropped, the transformational and variational explanations are not only compatible, they form the basis for a satisfactory approach to biological processes: A population changes as its constituent developmental systems change and perpetuate themselves with different frequencies.

Elliott Sober (1984:chap. 8) draws on Lewontin's distinction for his discussion of Darwinian natural selection, a *variational* process.[1] Some property of the variant acts as a "positive causal factor" in increasing its probability of appearing in the next generation. The focus is not on change in individuals but on *differences* among them. The individuals themselves may even be static. Lewontin (1982:155) says, "It is only the collection that changes, not the individuals of which it is composed."

Modern evolutionists, then, have rejected the transformational model of population change and embraced the variational one. This is often seen as the definitive victory of Darwinian over Lamarckian thought. Developmentalists have not been so absolute; they have tended to incorporate both models into their schemes. Operant behaviorism is an example of a selectional approach to ontogeny. Other examples are found in immunology and neurobiology (Edelman and Mountcastle 1978; Jerne 1967). Because of the gene-centered logic described above, however, the transformational model of ontogenetic unfolding predominates whenever a feature is thought to have an evolutionary history.[2] In fact, Lewontin (1983b) asserts that contemporary evolutionary thought has combined the Mendelian idea that internal factors make the organism with the Darwinian one that external factors shape the population. The opposition of internal to external causes is, as he points out, a deeply problematic one. Insisting on the primacy of one or the other for either ontogeny or phylogeny is not as useful, it would seem, as constructing an alternative to the framework that seems to demand a choice. An alternative is available, but it entails rethinking some of the metaphors that reflect (and support) this attempt to allocate causal efficacy to internal or external sources.

Although an emphasis on natural selection has sometimes promoted a static view of organisms (Gray 1988; Lewontin 1982; Sober 1984), it need not be so. A variational explanation need not preclude individual activity and development, and regular change in an entity need not be seen as the fulfillment of a plan. Moving beyond the alternatives, then, involves moving beyond the opposition of internal to external forces, toward a view of biological constancy and change as a function of interactive systems. I would like to examine the metaphors of selection by an agent, internal "programs" for innate traits, and heredity as trait transmission, and the ways each implies a kind of stasis. Interrelated as they are, they must change together if they are to change at all. Natural selection requires reliable life cycles, not static genetic programs or organisms. Ontogeny is the contingent functioning of entire developmental systems. It is the systems, mobile networks of organism-niche relations, that are "transmitted" (reconstructed) in heredity. These enlarged conceptions allow us to capture the relation between individual variability and population change without being committed to a static view of life.

Natural Selection: Action or Interaction?

SELECTION AS ACTION BY AN AGENT

The variational model of evolution, as we have seen, focuses on changes in the collection, not the individual. Sober (1984:149) asserts that selection, a variational process, permits and even requires a kind of stasis in individuals. Natural selection is a force external to the organism, "acting" on arrays of traits; change in the traits seems at best irrelevant to the process and at worst a threat to it. Why is this so?

The notion of choices made from a static array, first of all, encourages a momentary denial of development. In order to make decisions about biological "raw materials," the breeder on whom nature-as-selector is modeled typically surveys organisms in the same developmental state, so that they are as closely comparable as possible. If the adult character is of interest, developmentally earlier ones will be "invisible," at least until the adult state is predictable.

Conventional conceptions of heredity and development imply another kind of stasis. Traits under selection must be "inherited," and the use of that word to describe both population relationships and the ontogeny of certain phenotypic features encourages us to associate a sort of developmental stasis with a selectional history. Not only does the organism at selection tend to be treated as a fixed bundle of (fixed) traits; it tends to be seen as passive as well. But organisms choose and transform their environments even as they are affected by them. Selection is the outcome of certain interactions between organism and milieu, not the action of an agent on a passive object (P. P. G. Bateson 1988a; Gray 1988; Ho 1988a; Lewontin 1982; Taylor 1987).

SELECTION AS INTERACTION

These assumptions of stasis and passivity make the image of selection by the environment misleading in important ways.[3] Nature is not a deciding agent, standing outside organisms and waving them to the right or the left. However much we may speak of selection "operating" on populations, "molding" bodies and minds, when the metaphorical dust has settled, what we are referring to is still the cumulative result of particular life courses negotiated in particular circumstances. "Selection" is shorthand for certain kinds of changes in the distribution of interacting developmental systems. It refers to those population changes that occur when there is heritable variation (traits are differentially associated with lineages) and the variants interact with their environments in ways that confer on them different probabilities of being perpetuated. It is misleading to speak of a selecting *agent* at all.

As it is currently defined, evolution by natural selection occurs only when gene frequencies change. But one could be less restrictive and adopt a broader definition of evolution as change in developmental systems; the degree to which genetic change is involved would then be an open question. Lewontin (pers. comm., September 1986) uses a distinction made by Sober (1984): There is selection "for" properties that enhance the likelihood of survival and reproduction, and selection "of" the objects that recur in the next generation as a result. Lewontin suggests that selection is thus "for" an instantaneous state but "of" the developmental system.

As we come to understand the processes that produce transgenerational stability and change, we may continue to speak of natural selection because the concept has expanded with our understanding (e.g., to accommodate reciprocal influences between organism and environment), or we may decide the term is too closely tied to undesirable implications (stasis, action by an agent on a passive object) and choose another. Insofar as the former is true, the term will have lost much of its metaphorical power. Some would argue that metaphors do not so easily relinquish their hold on our thought and action, but what is finally important is how we construe the concept and how we demonstrate its appropriateness. The developmental system includes not only the organism but also the features of the extraorganismic environment that influence development. By positioning the organism in the ecological context in which it exists and grows, this view overcomes the alienation of organism and environment of which Lewontin has complained (1982 and elsewhere; see also Gray 1988) while restoring change and relation to the world we study.

Innateness: Persistent Essences or Reconstructed Systems?

PROGRAMS FOR NATURAL STATES

The concept of the innate is a second part of the reverberating circuit mentioned above. One of the many metaphors attached to it is that of the interior plan for development. Sober has traced the descent of the related idea of the "natural state" from Aristotelian essentialism. It appears in biology in several guises. Among them is the notion of a species essence, which will tend to be expressed as long as there is no interference; it is a "zero-force state" (Sober 1980). *Innateness* is unfortunately used to refer both to species-typical traits and to certain variable ones, so it is not surprising that individual variations can also be attributed to internal essences or plans. Thus a person's "genetic propensity" or "biological potential" is treated as that phenotype toward which he or she tends as long as extraneous influences do not intervene.[4]

Insofar as natural selection is said to act only on inherited traits, and insofar as these are treated as natural states represented in, and created by, the genes, selection implies a kind of developmental stasis. "Innate" characters are thought to be internally generated and trebly static: immutable in individuals, uniform across generations, and/or universal in individuals. They are attributed to genetic programs "for" those features, and are frequently assumed to be either inevitable or changeable only at heavy cost. Form and causal agency are placed in the nucleus, protected from commerce with the shifting world outside by Weismann's barrier (the segregation of germ cells from body cells in some organisms).[5] Thus the innate is opposed to the acquired as the necessary and timeless is opposed to the contingent and fluid. Although departures from the species or individual natural state are acknowledged, they are seen as *deviations* from the natural (even desirable) condition.

The importance of natural selection in evolution is a matter of dispute. (It may be less common than assumed; Sober's 1984 criteria for a positive causal factor seem very stringent; see also Hailman 1982.) But if a selectional history *is* indicated, it does not require developmental fixity (Lehrman 1970) or any other traditional marker of innateness.

PERSISTENCE AND RECONSTRUCTION

It is true that one can take the snapshot attitude mentioned above, stopping time to create arrays of features for comparison. Transgenerational stability can then give the impression of immobility; photographs of similar objects, even if widely separated in time, may seem to be images of the same object. We speak of "the shark," for instance, remaining "the same" for millions of years, as though it were a single shark, not a succession of life cycles. Interestingly enough, George Williams (1985:121), seeking to rebut Sober's criticism of his gene-level account of evolution, makes this very point: that phenotypes do not actually *persist* across generations, but rather *recur*. Genes, he says, persist. They are thus the real objects of selection. The obvious objection is that genes themselves recur by replication. This Williams counters by detaching the notion of the gene from its physical embodiment, characterizing it instead as a "weightless package" of information coded in the structure of DNA. But even leaving aside the complexities of overlapping and discontinuous genes and the problems they pose for the concept of coding, the information-as-sequence must be reconstituted as well. Disembodiment seems not to solve the problem of discontinuity. Williams seems, in fact, to be saying that physical persistence is not important after all. Reliable availability in each life cycle, however, is. Although I agree with him, I think this admission undermines his insistence on the primacy of the genetic level.

Williams (1985:121) writes of "natural selection *of information for its effects*" and says the logic of selection would permit other mechanisms of information transmission besides DNA replication. But if the particulars of gene replication are not necessary to the logic of natural selection, one could argue that evolution would proceed even if the genes materialized only a nanosecond before they were used. Although DNA does not behave this way, the chemical complexes for transcription and translation *are* assembled on the spot, just as many other aspects of the developmental system are. More to the point, if "information" is defined by its contribution to development, as Williams's phrase implies, one need not look far for alternative means of transmission; organisms have a variety of media at their disposal. Cell structure (both membrane and cytoplasm) is part of the developmental system that connects the generations, as are chemicals derived from maternal DNA and myriad extracellular and extraorganismic influences. Many of these are provided by parental physiology and behavior; others are reliable aspects of the larger environment, including the social environment. The "informational" function of any influence is determined by the role it plays in the developmental system as a whole. Regularity of gene function is thus a result of developmental regularity as well as a cause of it. Any biologically interesting notion of information must be interactively defined in this way, and what is crucial is not *permanence* but *availability at the appropriate time. Activity,* not stasis, and *relation,* not autonomy, are central to this conception. Persistence is beside the point in accounting for reliability.

It is possible for features to evolve without appearing in each generation or in all natural environments, and even the most reliable features have an ontogeny. If developmental courses are regular, they can for some purposes be taken for granted. (It does not follow that taking them for granted makes them regular, and taking them for granted is not the same as understanding them.) One can explain a bed of red flowers by pointing out that only certain kinds of bulbs were planted, even though bulbs have no flowers, the flowers are fleeting, and under other conditions the bulbs might have produced blooms of a different shade, or none at all.

Dynamic stability, then, is needed if natural selection is to affect a population. It is a contingent matter whether and how the conditions for construction of a feature are correlated with the conditions for its use. The variants must recur, and so must the causal background against which a given phenotypic difference makes for a difference in fitness. But, as we have already observed, people often reverse this reasoning and *assume* developmental stability from some evolutionary argument or other. Having concluded that a feature is in some sense adaptive, they may conclude that it evolved by natural selection (as traditionally defined, by genes), and thus tacitly assume this kind of long-term stability, even when the assumption may not be justified. Or they may project the stability into the future, predicting fixity without knowing whether the conditions it requires will be present (Kitcher 1985).

Heredity: Traits, Differences, or Systems?

Basic to much of the confusion about the relationship between evolution and development is the concept of heredity itself. The frequent conflation of population with individual levels is very much related to the dogma that evolution is fundamentally a matter of selection for genes that make organisms. Thus, emphasis is placed sometimes on *population* properties (variation or lack of it) and sometimes on *individual* phenotypic properties that are thought to reflect genetic causation (absence of learning, presence at birth, etc.). Genes that *go to fixation* in a population (become universal in it) are somehow assumed to *fix traits* in individuals; or genetically determined variance is thought to imply "genetically determined" traits.

We have seen that a variational mechanism permits stasis; Sober (1984: 149–152) claims that Darwinian selection *requires* stasis both in individuals and across generations. But I have argued that selected entities need not be static. What is required for stability across generations is heritable variation. Heritability is often thought to imply stasis of some sort (our third metaphor, the transmission of traits across generations, on the model of property inheritance), but what it really involves is *recurrence* in lineages. Offspring must resemble their relatives more than they resemble nonrelatives (Hailman 1982; Lewontin 1978; Sober 1984:151). This in turn involves dynamic regularity, which depends on order both outside and inside the organism. As reliable as these processes may be under some circumstances, their repetition is always a contingent matter, and though it is often assumed that variations in conditions must be large in order to deflect an evolved developmental course, small changes, especially outside the normal range, can have significant effects.

INDIVIDUALS: INHERITANCE OF TRAITS

What, then, is inherited? One seeking to come to terms with transgenerational regularity in living forms and functions may be tempted to employ the metaphor of hereditary transmission of traits discussed in the first chapter. One focuses on similarity between generations or on some aspect of the phenotype that justifies saying the feature was "passed on" from parent to offspring. An apparent benefit of this tack would seem to be that everyone "knows" what these words mean. Some things are in the genes in some unspecified (unspecifiable, really) way, while others are imposed from the outside. I have already indicated that the meanings that travel together under this rubric are neither logically required by each other nor reliably linked empirically. A trait that is present at birth, for instance, is not necessarily immune either to prenatal or postnatal variation in conditions (birth weight in humans is related to a variety of prenatal influences, and it does not always predict later size). A trait that involves no apparent learning is not always invariant within or between generations (insect caste membership can be induced by a variety of factors). A trait that shows parent-offspring similarity need not be longitudinally stable (a brunette and her dark-haired daughter may both have been blonde as infants). Reliance on this ready-made set of assumptions, then, is risky.

Because all aspects of the phenotype must develop, they are all "acquired" in ontogeny, whatever their distribution in the population. They are all "environmental" because particular conditions are required for their development, and because these conditions enter into the formation of the organism from the beginning. Phenotypes are all "inherited" and "biological" as well, if by this one means some sort of causal role of the genes in their development. If one seriously accepts the origin of phenotypes in causal interaction, as generally seems to be the case, judging by the literature, no distinction between inherited and acquired components of the organism is defensible.

POPULATIONS: INHERITANCE OF DIFFERENCES

When the statistical meaning of *heritability* is used, workers sometimes resort to locutions such as "inherited differences" or "innate variation" to refer to the relationship between genetic and phenotypic variance in the population. This has the appeal of being technically sophisticated. Referring to variation rather than to traits theoretically frees the user from unwanted nature-nurture implications. But it is quite hard to think of inheriting *differences* or *changes* (G. Bateson 1979:167–168). Gregory Bateson observes that a difference has no location; the term refers to a *relationship* (1979:109). It is even harder to know what it would mean to attribute selfishness to one (see P. P. G. Bateson 1986). The word *inherit,* after all, customarily refers to the passing of objects, land, money— roughly, *resources*—from one individual to another.

Heritability statistics depend on the constitution and distribution of developmental systems in a specific population at a specific time, and correlations among relatives may change. Human behavioral geneticists, in fact, may seek to document age changes in heritability (Plomin 1986). This relativity contradicts many of the implications of static, internal essence that inheritance has traditionally had (a good thing), but the developmental implications nevertheless tend to be attached to the variants in population discussions (not such a good thing).

INHERITANCE OF DEVELOPMENTAL SYSTEMS

I have suggested that we should think of heredity not as the transmission of *traits* between organisms, or even as the transmission of *differences,* but rather as the ways in which developmental *resources* or *means* become available to the next generation. A focus on process allows us to be true to the regularity these metaphors are meant to convey while recognizing the organized and organizing activity that creates both regularity and variety.

Traits do not pass from one organism to another like batons in a relay race. They must be constructed in ontogeny. This is true whether or not they are invariant in the population or have a traceable phylogenetic history. Most people respond to this obvious by truth asserting that what is *literally* transmitted is the genes (or information) that makes the traits, so the metaphor is innocuous. If, however, one believes that what is inherited is whatever it takes to make a trait, why be satisfied with a partial account? If one wishes, that is, to identify heredity with developmental *means* (the prerequisites for ontogenetic construction), one should include all of them; the organism then inherits (has available to it) a highly complex, widely ramified developmental system.

Some seek to solve the problem of the developmental insufficiency of the genes by including in the hereditary package some cytoplasm, transcripts of maternal genes, and other cellular features. This is marginally better than focusing on naked DNA. It tends to pass muster because it remains comfortably within the boundary of the egg, the traditional vehicle of heredity. But no egg can develop in a vacuum, and other ontogenetic requirements must be provided in some way. Many are reliably present in the wider environment, but many others may have to be supplied by the organism itself, by parents and other conspecifics, or even by organisms of other species. (Consider mutualisms, in which members of one species feed, protect, transport, or pollinate members of another. Systems can also be nested, as with endosymbionts.) The developmental system includes, then, not just genes, but whatever else in the living or nonliving environment contributes to or supports development. The system's constituents and configuration shift with its activity. In this view many parts of the changing complex are themselves generated by it. Hormones produced by the developing organism, for instance, serve as the chemical environment for further chemical interactions. Activity may provide necessary self-stimulation or social contacts. Whether the interaction is within the organism or between it and its milieu, both external and internal conditions are important, and the dynamic exchange characteristic of biological processes finally makes the division artificial.

If, on the other hand, what is inherited is that which makes one trait different from some other one (the "inherited differences" approach), *then not only the influence that makes the difference but the system in which it makes that difference must be inherited.* There are also many nongenetic influences that can play this role of difference maker, generation after generation (Cohen 1979; Gray 1988). Consider the ability of customs and social structure to maintain differences among humans. Whether one speaks of traits or differences, then, what one must ultimately reckon with is a full developmental system.

Development is thus basic to all notions of inheritance. What I have in mind is not the sort of unfolding of predestined fate involved in the transformational model. That would be revelation. Forms emerge. They are neither imposed from without nor "translated" from within. Nor do the genes give the basic outlines while the environment provides the details. Just as unsatisfactory is a dual model (some things unfold, some do not) or a continuum (there are degrees of programming). All development involves contingency and ordered (but real) change. It is a process generated by interaction of a heterogeneous mix of influences, and its mobile unity and order derive from its embeddedness in this matrix of causes, not from insulation against it.

Constraints and Possibilities

Sometimes the focus in heredity is not on what *is* generated, but on what may *not* be. Nature-nurture disputes are increasingly being recast in the language of constraints. But to see biological nature in terms of internal limits on extraneous developmental variation is simply to resurrect (albeit in attenuated form) the old concepts of natural states and developmental necessity that fueled the nature-nurture debate. At another level, scientists argue about developmental constraints on natural selection. Again, the concern is with what may not happen, and once more the opposition is between an internal resistance to change and an arbitrary external force.

In both cases the dispute is ultimately about the possibilities of developmental systems: about the ontogenetic possibilities of a genotype or the possibilities for variation in successive developmental systems. No rules, however, specify the range for either. There is no reason to suppose that all conceivable alterations are possible, or that all possible ones are equally probable. Nonrandom possibilities for change would seem to be part of our concept of a structured system. But such structure has a history; it comes into existence through activity and may change with further activity. It simultaneously generates change and directs or constrains it. *These are not opposing forces but two perspectives on dynamic order.* Like potentials, constraints are most usefully conceptualized as relational, not "endogenous," and as emerging in processes, not as prior to them. Possibilities for change evolve; they are generated in interaction. To oppose necessity (physical, biological, or developmental) to history, then, is to misrepresent both.

Developmental systems can vary in many ways. Sometimes a variation enhances the likelihood that the new system will recur. If the perpetuation of variations is a joint function of active organisms and surrounding conditions, and if development is structured but not predestined, then the variational and transformational accounts of population change can be synthesized (Gray 1988).

Biology and Culture

I have touched on some ways the environment has been personified in descriptions of natural selection. I have also maintained that the treatment of the "innate" as produced by static internal plans is more a product of conceptual confusion than an adequate empirical description. And I have examined the role of the metaphor of trait transmission in joining these views of evolution and development.

These mutually reinforcing metaphors often lead us to associate evolution with stasis. Stasis (or, better, some stable set of conditions and relations) is projected backward or forward in time. This is a paradox because evolution is defined by change, depends on change, and makes other kinds of change possible. Nevertheless, as I mentioned earlier, people infer developmental fixity from a phylogenetic history or assume a history of natural selection when a trait is pronounced "biological." Heredity does link development to evolution, of course, but the relation is misconstrued when the processes are misconstrued. I have proposed a wider definition of inheritance than is generally used, one based on process rather than product, on the "transmission" of developmental resources rather than of traits.

None of the assumptions of stasis associated with evolution by natural selection is justified. Stasis at selection is permitted by a selection mechanism but not required by it, even though our observational practices tend to freeze time in cross sections. The assumption that organisms under selection are passive reflects a tendency to think about these interactions as choices made by an outside agent. The stability requirement, that a selected entity must actually manifest itself in successive generations with some regularity, entails reliable ontogenetic reconstruction, and this is a matter of process, not immobility (Ho 1988a, 1988b).

NATURAL STATES, NATURAL CONCLUSIONS

The usual way to distinguish biology from culture is to attribute "innate" characters to genetic evolution, while "acquired" or learned ones are chalked up to "Lamarckian" cultural evolution. This is to rely on an old and deeply entrenched version of the nature-nurture opposition. If inheritance is defined in terms of developmental systems, no such distinction is necessary or even possible. Yet, even those who insist that developmental and evolutionary questions are not to be confused, who denounce the entire nature-nurture controversy as senseless, tend to associate the "biological" with deep, immutable truths, usually emotional or motivational ones, and the "cultural" with contingent detail, perhaps learned by imitation or conditioning. (However casually "imitation" is invoked in such discussions, the term embraces considerable diversity; see Galef 1988.)

The essentialist idea of a privileged developmental pathway and phenotype is very much alive in biology, psychology, and anthropology. Generally expressed in terms of biological "bases" or "propensities," it is a version of the natural state model described earlier. This is not what I mean when I advocate increased attention to development in evolutionary studies. If one believes in natural states and thinks they are properly termed "biological," perhaps it is natural to conclude that anything that has been dubbed "biological" represents one of these states (perhaps because it has been associated with a brain locus, occurs in some phylogenetic relative, appears to be difficult to influence in individuals, seems adaptive, etc.); in this thinking, it may thus be *natural* for a species or an individual. If humans are "naturally" polygynous, shared tendencies are at issue; if a person is a "natural" athlete, he or she is distinguished from the rest of us. One reads of "genotypic" aggression (Buss 1984; Konner 1982:203–207). Aggression, it would seem, can be in the nucleus, regardless of the behavior of the *organism*. Ultimate biological reality is somehow represented in the DNA, regardless of the outcome of any particular developmental course.

In addition to this vision of a hidden reality lurking in the genes, the natural state involves the idea of a tendency to approach that state. Sober observes that Aristotelian essences are "causal mechanisms of a certain sort" (1984:162). They are what we will be if nothing interferes. So we are told that we must carefully *"teach* our children altruism, for we cannot expect it to be part of their biological nature," a nature defined by their selfish genes (Dawkins 1976:150).

Allied to this notion of natural state is the assumption that such tendencies will be hard to deflect. Attempts to do so are likely to fail or to incur unwanted consequences. (Kitcher 1985:chap. 4 speaks of precluded states and states that preclude important desiderata. See Oyama 2000:chap. 6, for other examples.) Natural tendencies will press for expression, and we may suffer if they are denied. Instincts and psychological drives fit well here. We are even told what kinds of political systems are incompatible with our biological nature (E. O. Wilson 1978:190). Many social scientists have been quick to accept biologists' definition of a universal, fixed human nature and to draw conclusions about possible social arrangements (see, e.g., Peterson and Somit 1978). One question never addressed in these treatments is why only *deviation* from the program is thought to entail stress. This is a peculiar omission unless one assumes that the natural is also good and easy. This seems a small point until one realizes that inordinate focus on the putative costs of changing may divert attention from the costs of *not* changing.

The natural state model encourages a sequence of inferential leaps among the present, the past, and the future. It assumes just the conclusions we should be questioning and takes for granted just the complicated developmental and social systems we should be investigating. Finally, it draws attention away from the many kinds of regularity that would have been required as preconditions of such evolution.

Many questions about evolution, adaptiveness, or the role of genes in development are not only technical questions; they are also about the degree to which we feel we ought to view aspects of our lives as necessary, even good. When human customs and activities are at issue, the developmental stability we must assume extends far beyond what is required to produce recognizable morphology.[6] It even extends beyond "behavior," if by this we mean the relatively well defined patterns of movements observed in many other animals. Social life is made up not of *movements,* but of *acts,* which are largely defined by intent and meaning.[7] Intentions are construed, and meanings constructed, by agents interacting in a social-physical context that itself influences the significance of the interaction. We need a view of development that takes more seriously both the organism and its intimate interchanges with its surroundings (including its social surroundings). How do the nested systems work? A given child-rearing practice may have different meanings in different places or at different times, and thus different implications for the child's development and its relations with others (not to mention its reproductive fortunes). The human mind, in all its subtlety and variety, is social from its inception (a developmental psychology that is not social is no psychology at all).

UNIFIED SYSTEMS, MULTIPLE QUESTIONS

In a discussion of behavioral development, Gottlieb (1976) suggests some questions to guide research. How does experience facilitate, induce, or maintain development? Patrick Bateson (1983) adds a predisposing role, and points out that the same question could be asked of internal influences. Such questions undermine the attribution of internal necessity to some processes (and of "mere" external contingency to others) by some a priori criterion. At the same time, the questions direct attention to the causal factors that, in combination, could actually help us understand the phenomena that give rise to our concerns about developmental chance and necessity in the first place: degrees of reliability over time and across individuals. How do these constructions occur, and how do previous processes influence the likelihood of subsequent ones? How, in other words, do developmental possibilities and constraints arise?

If we move one level up, from individual ontogeny to what one might dub "Iterated and Interacting Developmental Systems" (companions to Gray's 1988 "Ecological and Evolutionary Cascades": IIDSs and EECs, pronounced, naturally, *ids* and *eeks*) that constitute evolution, we can ask similar questions, with perhaps similar benefits. To account for phenotypic regularity across generations we need to account for change and constancy in developmental resources and processes themselves, including those involving social interaction. For a feature to increase in frequency in a population, the conditions required for its development must be associated with increasing frequency. How are variations in developmental systems generated and perpetuated? Some perturbations will have no noticeable impact on an organism-niche system, while others will induce a change in it. In some cases the change will damp out in one or a few generations, while in others it will be maintained, and in still others, amplified. Other factors may predispose the system toward certain kinds of change or facilitate it when it occurs.[8] Genetic and nongenetic changes may be entrained. Gray (1988) observes that current research practice is not geared to discovering EECs. Laboratory researchers and breeders typically control conditions for successive generations. Such control may well minimize the probability of observing the kinds of novelties that might lead to a cascade. Whether one finds damping, maintenance, or amplification will depend to some extent on the source of perturbation or the nature of the change, but it will also depend on the available developmental and ecological resources—possible organism-niche complexes. The relationship between an influence and the rest of the system is crucial.

Focus on the ontogenetic construction of phenotypes eliminates the opposition between biology and culture. What we need is *not* ever more sophisticated ways to prize them apart, but rather a view of life and history that is rich enough to integrate the genetic, morphological, psychological, and social levels (each "biological," each with a history) in such a way that we are not tempted to indulge in phenotype partitioning at all.

In any developmental system, stability depends on the integration of lower levels into very broad ecological ones. For cultural stability, this means certain kinds of relationships among institutions, beliefs, values, and acts. It is wrong to assume that cultural facts can change quixotically while "biological" ones, however defined, cannot. If one is trying to determine whether a system is apt to continue to perpetuate itself, does it make more sense to extrapolate from past stability (or infer it from inappropriate evidence) or to look at the structure of the system itself?

ACTIVITY AND AGENCY

At a recent conference I watched a behavior geneticist diagram human development. Seeking to counter charges of oversimplification, she had filled a slide with arrows and loops. There were causes, influences, and moderators galore—but amid the maze of lines and labels, there was no *person* to be found. Just as evolutionists frequently regard organisms as evanescent epiphenomena to the real action in the gene pool (Dawkins 1976), so developmentalists, in their analytic zeal, have often ignored the unifying role of the individual, rendering it an abstract nexus of inner and outer causation. To see the organism-niche complex as both source and product of its own development is to acknowledge the role of activity in life processes—not genes organizing inert raw material into beings, and not environments shaping or selecting passive bodies and minds, but organisms assimilating, seeking, manipulating their worlds, even as they accommodate and respond to them.

To insist on the organism as a locus of agency is not to deny causality or to render causal investigation useless; it is not, that is, to embrace some concept of free will as uncaused cause. Nor is it to attribute human mentality to organisms that lack it. Still less is it to deny the social integration of individuals. It is, simply, to reinstate living beings in our analytic schemes. One benefit of this is that it makes clear where we must look if we wish to know what the possibilities for change are, for an individual or for the species: not at some set of disembodied constraints or rules or programs, but at relations within the organism and between it and its surround.

Biology is often contrasted with history, especially cultural history. Human history, however, is fully biological, not because it is predestined, but because it is the chronicle of the activities of living beings. Moreover, there is no "biological" character that does not have a developmental history. Missing from many views of cultural change and stability is the recognition that values, beliefs, and ways of behaving are ontogenetic constructions of a subtlety that is not captured by the metaphor of transmission, even "transmission by learning." The same is true of "genetically transmitted" characters. Much of cultural change occurs as individuals come to interpret old messages (including messages about biology and history) and practices in a new way; this in turn affects their impact on others. What comes of the chemical, mechanical, and social-psychological resources an organism inherits depends on the organism and its relations with the rest of the world. It makes its own present and prepares its future, never out of whole cloth, always with the means at hand, but often with the possibility of putting them together in novel ways.

5 Ontogeny and Phylogeny:

A Case of Meta-Recapitulation?

During the time I have been occupied with the nature-nurture opposition, I have become sensitized both to the various guises in which this dichotomy appears and to structurally similar ones in other fields (figure 2). One of the sources of the nature-nurture dichotomy in science is the field of epistemology, in which a classic problem has been the origin of knowledge. The disputes between rationalists, who insisted on innate ideas as the source of knowledge, and empiricists, who credited the senses, did much to set the framework for more recent disputes. Similar oppositions are found in other fields as well.

FIGURE 2. *A sampling, from a variety of fields, of dichotomies in which internal factors are opposed to external ones.*

Selected Dichotomies

Epistemology	: innate vs. experiential sources of knowledge
History	: internalist vs. externalist explanations of change
Anthropology	: biology vs. culture
	: culture vs. geography
Biology	: history vs. ecology (ecology)
	: physical necessity vs. chance (origin of life)
	: mosaic vs. regulative development (embryology)

Psychology, biology, and contemporary philosophy inherited the epistemological question fairly directly, but a number of other contrasting pairs have less obvious ties to the nature-nurture complex per se. Among historians, for instance, one finds internalist and externalist views of historical change, including scientific change. Anthropologists attribute cultural patterns in a society to biology or to culture (a form of nature-nurture dichotomy), and, interestingly enough, argue about cultural versus geographical determinism as well (Netting 1978). What is worth noting in the two anthropological examples is that culture is treated as an external factor in the first case and as a conservative, internal one in the second. This suggests that it is often scale (whether of time or magnitude) or level of analysis that is at issue in these debates, not competing factors of the same type.

Russell Gray (1989) has discussed ecologists' pitting of historical (phylogenetic) factors against ecological ones in explaining relationships between populations and their niches. This resembles the anthropologists' opposition between culture and geography, which contrasts "inherited" ways of life with the immediate conditions of life. The two disciplines tend to rely on different senses of *inheritance,* of course, and they use somewhat different time scales, but in both cases, history moves "inside," playing the role of a stabilizing counterforce to the immediate demands of the environment. There are other biological oppositions, too. Students of the origin of life emphasize either physical necessity or chance (Ho and Fox 1988). Embryology, meanwhile, has had its disputes over mosaic and regulative development. In the former, tissues seem "fated" to differentiate in a particular way and do so even if transplanted to different locations in the developing organism. In regulative development, by contrast, subsequent differentiation of transplanted tissues is influenced by their new location, so a normal result is obtained. The embryological example is distinctive in that developmental regulation traditionally implies responsiveness of developing tissue to its immediate environment but involves the production of *typical* outcomes, not variant ones. In other inside-outside oppositions, however, such sensitivity to local conditions tends to be associated instead with variant outcomes. Once again it seems that differences in level of analysis—in this case between tissues and organisms—are crucial.

There are varying degrees of similarity among these oppositions. Tracing their histories and relations would be a fascinating exercise in itself. Instead, however, I would like to examine some striking resemblances between the nature-nurture dispute in developmental studies and the argument over developmental constraints and natural selection as competing explanations of evolution. The sorts of conceptual problems that are generated in the first may also arise in the second (hence the "meta-recapitulation" of the chapter title). The view of development I have been elaborating eliminates the need for the developmental dualism of the nature-nurture opposition; here I sketch a related formulation of evolutionary change that calls into question the opposition between constraint and selection as well.

Consistency and clarity are, in my view, reason enough for the kind of analysis engaged in here, but there are other reasons as well. Conceptual problems have had significant consequences in the nature-nurture debate. When ambiguous terminology leads to misunderstandings, even serious ones, the remedy appears to be straightforward: Be a bit more careful with your words. Indeed, one may be chided for quibbling over what things are called. Usually, however, the matter is not so simple. Conflations that are persistent and recurrent, as they are in the nature-nurture opposition, can indicate serious difficulties with the reasoning itself. In addition, I am convinced that it is useful to examine the often subtle assumptions embedded in theory. These assumptions about causality, agency, and process are not only important for scientists, they are also intimately related to our views of our own actions and possibilities.

The question mark in my chapter title is a real one. While I have devoted considerable time to documenting the mischief made by the nature-nurture dichotomy, it is less than obvious that the constraints-selection distinction necessarily involves difficulties of the same variety and magnitude. This chapter is an invitation to biologists and philosophers to consider just how far the parallels between the two disputes extend. The disanalogies may be as significant as the analogies, and disputes over evolutionary dynamics, unlike arguments about innate selfishness or aggressiveness, seem rather removed from everyday matters, appearing not to have the same sorts of complex implications. Yet, as I will point out later, the two discourses, the developmental and the evolutionary, are not independent.

Parallels

An important question in developmental studies, if not *the* important question, pertains to the ontogenetic origin of organismic form and function, including the form of the mind. Traditionally, answers have focused either on a set of basic structures supposed to be transmitted in the genetic material or on the contingencies of individual experience.[1] Thus, nature-nurture battles are ostensibly about the *allocation of causal responsibility* for development either to the genes or to the environment. The motivating concern, however, often appears to be with some notion of *limits* rather than with actual details of causation. What people finally seem to be speculating about is the limits on possible phenotypic variation and change.

The prime question in evolutionary theory is also about the origin of form, but over phylogenetic, not ontogenetic, time: Internal developmental constraints are set against the contingencies of natural selection. Here again, the questions are couched in terms of the *relative influence of alternative causes,* and again the basic concern seems to be with limits, but this time the limits on phylogenetic variability and change. In both fields, one finds a kind of causal dualism, then, as internal forces are opposed to external ones. I submit that these quarrels over causal responsibility miss the point, and that the point in each case is developmental dynamics: the possibilities for alternative developmental outcomes in a single lifetime (Could this person have developed differently? How differently?) and the possibilities for change in developmental systems across generations (In what ways and how much can this lineage change over evolutionary time?).

A somewhat more detailed account of these parallels follows, but there is also a more direct relationship between the two dichotomies (figure 3): Ontogeny appears in the evolutionary debate as a constraint on evolutionary change, and the internal, genetic, factors in ontogeny are the legacy of that evolutionary change. Since the opposition between nature and nurture is so intimately related to the one between constraints and selection, it seems natural to wonder whether their resolutions may be related as well.

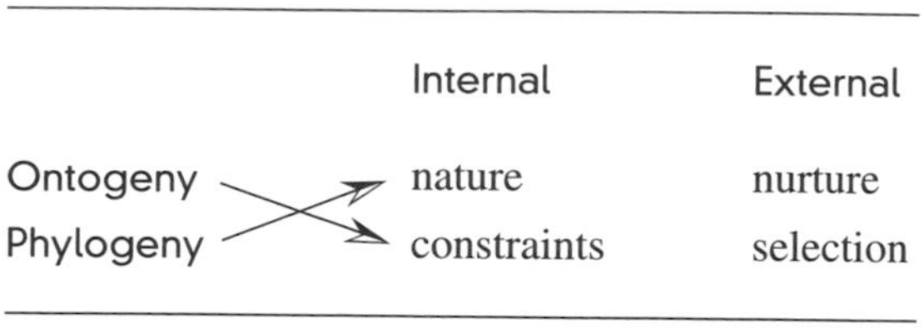

	Internal	External
Ontogeny	nature	nurture
Phylogeny	constraints	selection

FIGURE 3. *Mutually supporting dichotomies in development and evolution.*

Of the many points that could be made here I wish to highlight just four. (Three of these are sketched out in Oyama 1992, and Gray gives more extensive treatments in his 1987a, 1988, and 1989.) In both debates, internal causes are contrasted with external ones; in both, fixity is associated with the former and malleability with the latter; and in both there has been considerable oscillation between internalist and externalist perspectives. Finally, similar compromises have been proposed, as theorists attempt to reconcile internal to external factors. These are considered in turn.

Insides and Outsides

The nature-nurture debate turns on a separation of insides from outsides. So does the debate about constraints and selection. As we saw in chapter 4, Lewontin (1982, 1983a; see also Levins and Lewontin 1985) and Sober (1984, 1985) have elaborated two models of change that offer a convenient way of seeing how this occurs in each case. In the transformational model, change in a collection results from change in the constituent entities, and the entities change in a uniform and predetermined manner. Recall that in the variational model (which includes selectional processes), variant entities are propagated with differing frequencies, and the variants themselves are static. Change is generated from within in the transformational model and imposed from without in the variational model (figure 4).

FIGURE 4. *Two models of change. Adapted from Lewontin 1982, 1983a; and Sober 1984, 1985.*

Transformational:	1. Change in a collection results from change in the constituent entities.
	2. The entities change in a uniform and predetermined manner.
Variational:	1. Variant entities are propagated with differing frequencies.
	2. The variants themselves are static.

The transformational model dominates in developmental theory. Maturation, as it is traditionally conceived, is the quintessential internally driven process.[2] It is to capture the regularity of such ontogenetic sequences that metaphors like the genetic program are invoked: Biological "nature" is treated as both the cause and the effect of these regular processes. The "program" that begins by describing the predictability of normal development thus becomes a quasi-explanatory device as the reliable reappearance of certain features in successive generations is attributed to central control. Species-typical processes and characters constitute the fixed core that is transmitted in the genetic material.[3]

The dominance of the transformational model of development is such that even the most committed believer in the importance of environmental influences accepts a genetically given body and a set of reflexes, instincts, or some other substrate for behavioral development. An inherited structure, with all the preformationist assumptions this entails, is a prerequisite for all of the most common sorts of nature-nurture haggling. Disagreements, then, occur against a consensual background of intrinsically driven processes; indeed, developmental studies are virtually defined by the transformational model (Morss 1990). Sober (1984), in fact, refers to the model of internally driven change as "developmental" rather than "transformational."

Nurture in these exchanges over developmental causation is not so unitary. Selection, a variational process, is frequently contrasted with instruction, for instance, but both are supposed to be elaborations of a pregiven "biological base," and both are seen as externally directed. Selectionist explanations are found in the literatures of operant conditioning, neurogenesis, the immune system, and cognitive and linguistic development.[4]

If the internalist perspective defines developmental studies, the field of evolution presents a rather different picture. Variational processes, including genetic drift and natural selection, dominate the modern synthetic theory. Populations are "molded" by these external forces. Orthogenesis, a transformational process, was resoundingly rejected by the Modern Synthesis. The virtual disappearance of this internalist, "developmental" perspective from evolutionary theory is generally seen as the triumph of scientific Darwinism over mystical Lamarckian progressivism.[5] Effectively abolished from contemporary evolutionary discourse (but see Grehan and Ainsworth 1985), *orthogenesis* is probably too heavily freighted a term to gain serious attention now. Its internalist connotations persist, however, in the literature on developmental constraints, a literature that is gaining in volume and influence. A notable attempt to integrate its diverse critical and empirical viewpoints into mainstream neo-Darwinism is evident in the group report on a conference convened for just that purpose (Maynard Smith et al. 1985). In addition to presenting a typology of constraints, this paper repeatedly points out the importance, and difficulty, of separating internal factors from external ones.

Malleability and Fixity

Variational processes are secondary to transformational ones in conceptualizing ontogeny, then, while the reverse is true in conventional thinking about phylogeny: Natural selection tends to be seen as the major form giver in evolution, while developmental constraints serve merely to narrow its scope. This separation of insides from outsides brings us to our second parallel between the developmental and evolutionary dichotomies: In both, malleability is associated with external factors, and fixity with internal ones. One even finds the same vocabulary used to express these relations. "Shaping" is frequently considered by both developmentalists and evolutionists to be capriciously variable; it represents the vagaries of individual or species history. Shaping and molding, in fact, are probably the most common metaphors for natural selection, conjuring the image of an omnipotent, omnipresent hand and eye (and will), the coordinated apparatus of the artisan. The shaping metaphor obscures the difference between the origin of variants and their perpetuation, a difference emphasized by many developmental constraints theorists.

Shaping is also a technical term in psychology's operant theory, referring to the gradual production of a complex behavior pattern that does not occur spontaneously. Behaviorists, in fact, have explicitly compared operant shaping with biological evolution (Herrnstein 1989; Skinner 1981). Successive approximations of the target behavior are reinforced in shaping, often by a trainer who progressively raises the performance standards; the problem and the criteria for reinforcement are thus set by the environment, though not necessarily by a human trainer.

In both cases an emphasis on external shaping involves a tendency to take the generation of the selected variants for granted, to view it as unproblematic: unsystematic and random, at least with respect to selection. In the same way that the context of scientific discovery has traditionally been considered to be independent of, and irrelevant to, the real business of justification, the mechanics of the generation of variation is treated as irrelevant to natural selection, as long as there is heritability. To focus on constraints, however, is to ask why some variants arise and not others, and this question tends to be conceived as involving internal causes. Thus one hears of genetic constraints on learning (species differences, often invoked to explain failure of conditioning; see Hinde and Stevenson-Hinde 1973; Shettleworth 1972) or of the genes limiting possible social arrangements (E. O. Wilson 1978).

Although any genotype is the result of an evolutionary history, in developmental studies the genotype frequently plays the role of an enduring essence insulated from change, the unmoved mover that both *embodies* a plan for an organism and *executes* it. In similar manner, developmental constraints on natural selection are sometimes presented as timeless physical laws or ahistorical necessity.[6] They limit the otherwise untrammeled variation stemming from environmental fluctuations, although a distinction may be made between universal constraints and those affecting only some evolutionary lines (Maynard Smith et al. 1985). When constraints are local or historical, they constitute a genealogical legacy. They carry the lineage history and influence its further history; they are species "nature" to evolutionary "nurture," and are thus more like the two oppositions in figure 2 in which past history is treated as an internal brake on current change: cultural versus geographical determinism (in anthropology), and evolutionary history versus ecology (in biology). Whether they are universal or lineage specific, developmental constraints tend to be conceptualized as placing *limits* on the arbitrary action of selection, forbidding certain forms and permitting others.

In both the developmental and the evolutionary literatures, then, *resistance* to change is more salient than the *direction* or *generation* of change, although as I pointed out in chapter 4, these are simply two aspects of biological dynamics. In fact, the language of direction and guidance can be used to express the same phenomena that are at other times described in terms of constraints: Behavior theorists sometimes speak of "predispositions" to learn, and constraints theorists refer to "biases" and "propensities." This dual role of constraints (resisting and directing) does not go unnoticed (Alberch 1980; S. J. Gould 1989b; Maynard Smith et al. 1985). (The subtitle of Hinde and Stevenson-Hinde 1973, on constraints on learning, is "Limitations and Predispositions.") To emphasize the orderly generation of variants is to give internal factors a more prominent, creative role, while external factors may then be cast as secondary filters. A constraint, as Stearns (1986:23) notes, is usually "imported from outside the local context to explain the limits on the patterns observed." S. J. Gould (1989b:516) observes that the term is a theory-laden one indicating that the causes being identified are other than the "canonical causes" in the theory.

Although Gould makes a case for retaining the term and enlarging its meaning, it is not surprising that *constraint* tends to be supplanted by other terms when researchers are making a bid to move their tradition from the periphery to the center. Grehan and Ainsworth (1985), for instance, seem to take issue with the tendency of the constraints literature to emphasize limitation; their position is that selection is "subsidiary" to orthogenetic tendencies to vary in certain directions. Similarly, Goodwin (1982a:112) urges a "rewriting of the origin of species so that 'origin' is understood primarily in its logical, generative sense, and secondarily in historical terms"; and Ho and Saunders (1979:589–590) "place more emphasis on the physiological and developmental potential (or internal factors) of the organisms as opposed to the 'external' factors of random mutation and natural selection."

The language of internal resistance to change (and of internal tendencies channeling change) brings to mind the image of recalcitrant material in the hands of an artisan; it recalls an ancient distinction between matter and form. The artist's freedom is limited by the nature of the material, and the notion of raw material with its own stubbornly causal properties is present in these contemporary debates about ontogenetic and phylogenetic change.[7] Because fixity is attributed to internal causes, and malleability to external ones, queries that are unintelligible if taken literally (whether internal or external factors are responsible for some feature) may be seen as (admittedly odd) ways of asking how amenable an organism is to various developmental influences, or how susceptible a species' developmental processes are to various kinds of transgenerational change. Living beings, however, are not made of static stuff, so it is dynamic stability that must be explained, and such stability is attained and maintained by constructive interaction, not isolation.

Oscillation

The third parallel between the developmental and evolutionary debates is that both have been characterized by an oscillation of received wisdom between the internal and external poles. Early psychology's preoccupation with internal entities such as minds, drives, and instincts led to the behaviorist obsession with external causation. This in turn led to the resurgence of "biological" approaches (behavior genetics, physiological and evolutionary studies), typically understood as focusing on "intrinsic factors." (Notice again the associations among biology, internality, and fixity.) Certain fields then witnessed a partial retreat from strict "programming" accounts. Selectional theories of neurogenesis, cognition, and the immune system were mentioned in chapter 4. These theories tend to be linked to very traditional maturational stories, though; selectional theories of language acquisition and cognitive development are, in fact, strongly nativist, both in intellectual lineage and in current orientation.

My earlier point about the preeminence of transformational explanations in developmental studies is relevant here; these research trends do not signal a different way of conceptualizing development in general. Rather, they are sophisticated efforts to reconcile traditional maturational explanations with certain empirical challenges: the astonishingly specific responsiveness of the immune system, for instance, or the obvious necessity for a theory of language development to accommodate differences among languages.

In evolutionary theory, the current interest in internal constraints seems largely to be a reaction against the hegemony of natural selectionist explanations, while, as already noted, neo-Darwinist selectionism is often seen as the refutation of earlier, more transformational visions of species change. Predictably, perhaps, some sociobiologists, whose power to explain the world depends heavily on the "power" of natural selection (an idiom we shall return to), have responded to the constraints literature by reaffirming their faith in the omnipotence of selection to mold bodies and minds (Noonan 1987; Thornhill and Thornhill 1987).

Compromises

Not only does one find oscillation between poles in both the developmental and the evolutionary debates, but the same kinds of (largely unsatisfactory) compromises described in chapter 3 are encountered as well. Developmentalists have often "solved" the nature-nurture problem by dividing the territory: by attributing some features to genetic control and others to environmental influences. But if there are fundamental conceptual difficulties with the very notion of genetically or environmentally directed development, it can scarcely be an improvement to combine them in some odd organic patchwork.

Apparently more judicious is the attempt to *quantify* the relative amounts of genetic and environmental influence on various features. There have been many important critiques of these strategies of phenotype partitioning, as well as of the partitioning of variance associated with such statistical techniques as the analysis of variance (ANOVA); even the seemingly more promising tack of tracing phenotypic "information" to the genes and the environment is problematic.[8] One difficulty with the partitioning of variance is that it frequently slides into the partitioning of phenotypes. Another is that, even if the eye is fixed firmly on variance, and not on variants, the results of local analyses tend to be confused with more general functional relationships (Lewontin 1985). A third, related difficulty is the ambiguity of terms such as *genetic control,* which are used to explain both invariance (inevitability or species-typicality, for instance) and some variance (genotype-associated phenotypic variation in ANOVA). Such usage encourages exactly the erroneous inference of developmental fixity from heritability coefficients that has plagued behavior genetics for decades.

Largely in an attempt to avoid such interpretational difficulties, some theorists have tried another tack: "genetic imperialism," in which genes are said to determine the range of possibilities while the environment selects the particular value (see chapter 3). In this strategy, the genes are given higher-level, generalized control, while secondary influence is doled out to nongenetic factors.[9]

It is not accidental that both partitioning and imperialistic conquest are typical outcomes of geopolitical conflict. Nor do I use these terms unreflectively. Indeed, I see nature-nurture arguments as territorial disputes in which the contenders strive to retain as much power as they can (to explain, to control what observations are counted as data, to dominate a field of inquiry, to subsume related areas), even as they make necessary concessions. Nor do I think that issues of "turf" are trivial or devoid of intellectual significance, although they are often presented that way. They represent entire traditions of theory and research, which in turn represent conceptions of science, of scientists, of knowledge and the world.[10]

As the reader has no doubt realized, these impulses to compromise are detectable in the evolutionary literature as well. When theorists attribute a character to selection or to constraints, they are implying that species features can be credited to alternative formative factors. Asking how much a feature owes to each factor, meanwhile, is analogous to quantifying genetic and environmental contributions to some part of the phenotype.[11] When, in commenting on this literature, Thomson (1985) wonders whether giving a developmental account of a feature is enough to eliminate a selective account, he seems to be expressing misgivings about this alternative-causes assumption. Trying to separate causes by partitioning variance into selective and constraining components, furthermore, presumably involves at least some of the difficulties that attend ANOVA in behavioral studies. (See, e.g., Stearns 1986; Thomson 1985; for discussion, see Gray 1989.)

Sober has analyzed attempts to apportion causal responsibility to the genes and the environment. He concludes that questions about relative causal contributions to the phenotype are not locally answerable because an organism cannot be affected by one and not the other. One can, however, ask how much difference one or the other made to a particular *set* of outcomes (Sober 1988). There is a distinct possibility that the same is true of attempts to treat constraints and natural selection as competing forces. Elsewhere, Sober (1987) discusses disputes over the relative strength of natural selection and correlated characters (that is, characters whose evolutionary alterations are linked because they are developmentally linked). When he asserts that the disagreement is not really about the power of natural selection (which is the way it is often cast), but rather about the "power of mutation," he seems to be identifying the same confusion between local and nonlocal analyses as exists in nature-nurture arguments. What appear to be questions about relative causal contributions to a *particular* outcome are intelligible only if recast as questions about a *class* of outcomes. He suggests that one could ask "*how often* a trait is maintained by pleiotropy [multiple effects of a gene] even though it is selected against, or is eliminated by pleiotropy even though there was selection for it. This is a far cry from looking at a single trait whose presence in a population is the joint product of selection and pleiotropy and asking which contributed more to its evolution" (pp. 115–116). Interestingly enough, Stearns (1986) asserts that selection and constraint are involved in all evolution; the problem is to determine the relative influence of external and internal factors. Behavioral scientists justify their continued pursuit of genetic and environmental "components" in precisely the same way.

Above, Sober (1987:116) refers to the "power of mutation." "Mutation" can refer both to alterations in DNA sequences and to the phenotypic consequences of such alterations. Restrictions on the range of phenotypic results of genetic mutations surely involve not just constraints on DNA changes but constraints on the rest of the developmental systems in which they occur as well. A particular DNA change may have no effect in some systems, and a variety of effects in others. The outcome will depend both on the alteration and on the rest of the system. It might thus be more apposite to avoid speaking of the power of natural selection or mutation, and to speak of *possibilities for variation in developmental systems* instead. Whether any particular phenotypic variation will occur is a function of the system dynamics, and it is the dependence of such variation on this interactive complex that is indexed (but not captured) by the notion of mutational power.

It might be wondered (and has been, by Kim Sterelny, pers. comm., August 1990) whether this is just another version of the imperialistic move. The short answer is "no." A slightly longer one is: "Sort of, but not really." Genetic imperialism seeks to decontextualize gene action by collapsing all possible ontogenetic outcomes into some notion of "genetic information," while the developmental systems formulation makes contextual dependence explicit by stressing the joint determination of outcome by the system and by its perturbation. The real imperialistic move for a developmental systems theorist would be to claim that a system "determines" all of its possible changes prior to specification of the particular perturbation.[12]

Any outcome of a multiplicative function is specified by one factor, *given the other one(s)*. Genetic imperialism gives the genes the power to specify all outcomes *given only themselves*. This is like saying that the number 2 "specifies" the products of all multiplications in which it might possibly be a multiplier, and that it does so before the fact; the multiplicand simply selects from this prior array—a most peculiar claim. Read for its rhetorical function, genetic imperialism can be seen as a ploy that makes certain causes recede into the background. Developmental systems block this move for either "internal" or "external" factors by including them both. If arbitrary causal domination is abolished within a system's boundaries, is it still an empire?

Just as developmentalists have sometimes shunned partitioning in favor of the range-of-possibilities compromise, so some constraints theorists have spoken of developmental laws determining the range of forms attainable by evolution. Gene changes simply select from this array (see note 12). Treating constraints and selection as competing causes implies that they can be separated. As Dyke and Depew (1988:117) point out, invoking constraints requires a baseline of pure selection. But "if the novelty generating process and the selection process are coupled and interactive" rather than independent, no baseline of "pure selection" can be described. By the same token, it would be difficult to describe an array of possible forms prior to, and independent of, selection.

Interacting Dualisms, Interactive Systems

Earlier in this chapter I pointed out that the nature-nurture and constraints-selection dichotomies are not only structurally similar, they are related in substance as well. In fact, they are mutually reinforcing. Evolutionary theory maintains developmental dualism by the logic described in chapters 1 and 3: Evolution is typically defined by change in gene frequencies. Organisms must therefore be explicable in terms of genetic "information." Genetic transmission is thus the needle's eye through which innate characters must pass. In Bonner's (1974) arresting image, the life cycle must (often) narrow to a single cell,[13] so inherited "nature" must be passed on in the zygotic DNA. Acquired characters, those formed by environmental action, are excluded and thus rendered inconsequential to evolutionary change. So evolutionary theory seems to require developmental dualism. Developmental theory, in turn, insofar as it embraces the transformational model of ontogeny (as predetermined, uniform, and internally generated), reinforces dualism in evolutionary studies by legitimizing this vision of autonomous change.

Resolving the nature-nurture dichotomy involves, ironically, taking development seriously. In a discussion of dichotomies in biology, Gray (1989; see also P. Bateson 1983; Johnston 1987; and Lehrman 1953, 1970) suggests that this resolution has been achieved in studies of behavioral development, but I think his statement should be taken less as an accurate report from the front than as a rhetorical device to prod evolutionists into action. (What biologist, after all, would be indifferent to having psychologists held up as models of intellectual sophistication?) I am less sanguine. One of the many factors conspiring to maintain, and repeatedly reconstruct, the nature-nurture complex is psychologists' increasing attention to evolutionary biology. Although the distinction between inherited and acquired traits that evolutionists want to make can be managed without any specific assumptions about the nature of developmental processes, it seldom is; in fact, it is typically treated as a statement *about* developmental mechanism.

The metaphors of transmission and programmed development, then, link ontogeny to phylogeny: Innate characters, fashioned by natural selection, are passed on in the genes and so reappear in each generation. "Transmission" is a metaphor for this reliable reappearance in each generation. What the image of biological faxing elides is the multitude of interactive changes over time that constitute epigenetic emergence and stability. The technical language of transmission genetics is correctly employed for particular distributions of developmental products (phenotypes) in populations, *but reliable developmental courses are the prerequisites for such population patterns.* Those life courses are *assumed* by the transmission metaphor, not *explained* by it.

The alternative I have offered for these nondevelopmental notions is the metaphor of construction. Emphasis on constructive interaction defuses the necessity of attributing power either to external artificers or to autonomous internal forces. This view of ontogeny may prepare the way for an evolutionary theory that is synthetic in a rather different sense from the usual one. It may allow, that is, the fusion of the transformational and variational models (Gray 1988; Levins and Lewontin 1985; Oyama 1992).

Synthesis

In the synthesis offered by the notion of the evolving developmental system, nature and nurture are no longer alternative causes; they are developmental *products* (natures) and the *processes* (nurture) by which they come into being. Since ontogeny is not an autonomous transformational process, it is not useful to think of developmental constraints as insulated, internal, and necessary. And since natural selection is not an agent choosing or shaping passive organisms, but is instead the result of the "interpenetration of organism and environment,"[14] natural selection is better conceptualized as something other than an arbitrary external force. Since neither model is adequate to explain change, furthermore, the partitioning compromise discussed above is again called into question.

Instead of opposing transformational and variational accounts, or apportioning causal responsibility between them, or giving one the role of determining the range of possibilities while the other selects from that range, we can return to the models themselves (figure 4) and dispense with the problematic aspect of each. Change in a collection can certainly result from change in its constituent "entities" (developmental systems), but the change need not be uniform or predetermined. Variant systems can propagate themselves with differing frequencies, but the variants themselves need not be static. (It should be noted that developmental systems are not cleanly bounded entities, and they never reproduce themselves precisely.) A more ample view of development leads to the more ample definition of evolution introduced earlier in this volume: Evolution is change in the constitution and distribution of developmental systems, organism-environment complexes that change over both ontogenetic and phylogenetic time.[15]

My ambivalence about the internal constraints literature resembles the discomfort I feel about many aspects of the "biological" trend in psychology: I am wary of a pendulum swing along a dimension I consider inappropriate. If similar conceptual problems exist in the constraints-selection literature, similar inferential traps may be present as well. The scholar who is alert to this possibility might be cautious about several things: Is quantitative language being used about a particular taxon when it is appropriate only for patterns in an array of taxa? Are local analyses—of character variation across taxa, for instance—being inappropriately used to indicate general functions? If terms like *constraints* are being used in several senses, are the senses made explicit and consistently distinguished? (Even the review by Maynard Smith et al. 1985 details a number of meanings.) If the various senses are conflated, does this lead to evidence for one kind of constraint improperly being used to infer another, in what I have called "cross inference"? Are factors that are interdependent and changing treated as independent and static? And are certain kinds of population-niche interactions being slighted as a result? Some biologists are now examining the methodological strategies of traditional approaches and exploring alternative ones (Dwyer 1984; Gray 1987a, 1987b; Levins and Lewontin 1985; Lewontin, Rose, and Kamin 1984; Patten 1982; Taylor 1987).

Evolutionary fixity does not carry the political and moral implications that frequently accompany questions of developmental fixity, so hasty conclusions about constraints on natural selection would seem to be less sociopolitically mischievous than are ill-considered pronouncements on human nature (or on nature in general). But given the interrelations of the two dichotomies, perhaps we should not be too complacent. Much of the controversy over the proper scope of natural selectionist explanation, after all, has been stoked by dismay over sociobiologists' claims about human mentality and culture. The assertion that some behavior or institution has an evolutionary explanation rather than a social one (or vice versa) feeds right into the assumptions in question (Oyama 1989). Herrnstein (1989:40) asserts that "nature versus nurture in regard to behaviour is the last great evolutionary controversy." Greater circumspection about such arguments may be in order.

If development is to reenter evolutionary theory, it should be development that integrates genes into organisms, and organisms into the many levels of the environment that enter into their ontogenetic construction. Explicit inclusion of these developmentally relevant factors, so often marginalized in conventional accounts, makes clear the context dependence of questions about developmental timing, universality, immutability, spontaneity, and other "biological" characteristics. It also makes clear their heterogeneity: There are many kinds of biological argument, and they should not be discussed in a manner that lumps them together willy-nilly. We can investigate questions about evolutionary history, current reproductive or survival advantage, development, and the causal processes by which behavior occurs (and explore the relationships among these questions) only if we first distinguish them from each other (Tinbergen 1963).

The view of change presented here is systemic and interactively constructivist. The unidirectional causation of the transformational model can sometimes be found in circumscribed analyses in which a single factor is isolated by controlling everything else, but the results of those analyses must eventually be reintegrated into a larger framework (chapter 6). When Bonner speaks of the life cycle narrowing to a single cell, he is making an important point about vital continuity, but even the zygote requires, and reliably has, the rest of a developmental system: all the developmentally relevant aspects of its world. The rest of the system shows continuity just as surely as the genes do, and as is the case for genes, this is often achieved through reconstructed structure and function, not through static material identity (not that "material identity" is a straightforward matter, especially if one takes a fine-grained approach). The continuity of the germ line, that is, is achieved by repeated reconstruction of DNA strands, and the continuity of other developmental interactants may also involve reconstruction rather than simple persistence.

Study of the life cycle includes the investigation of these many influences and of the ways they become (or do not become) available at the time they are needed. In organisms that reproduce in this way, the nucleus, cytoplasm, membrane, and the rest of the germ cell can pass through the eye of the reproductive needle—evolution's narrowed eye—but the other interactants need not; what is important is that they be reliably present in the next life cycle.

Resolution of the nature-nurture dichotomy is, I am convinced, good in itself. In this sense, I agree with those theorists whose attempts at compromise and resolution I have criticized here: Something was amiss. In addition, however, this resolution may also make us hesitate before pitting internal constraints against external selection when we conceptualize the emergence and persistence of form in evolution. That is, it may reduce the danger of recapitulating the sterile debate over inner and outer causes in our thinking about stability and change in evolution.

6 The Accidental Chordate:

Contingency in Developmental Systems

The somewhat peculiar title of this chapter is not just an allusion to items of 1990s popular culture. It refers as well to Steven Jay Gould's (1989a) book *Wonderful Life,* in which he discusses the fossils of the Burgess Shale, a rich bed of paleontological remains in Canada. Many of these fossils belong to extinct phyla. An extinct species or genus is hardly remarkable, but the disappearance of groups as high in the evolutionary hierarchy as the phylum may be, retrospectively, at least, more arresting: Consider that the vertebrates, as varied as they are, are only a subdivision of the phylum Chordata.

Gould enlists the bizarre Burgess creatures, which he calls "weird wonders," in the service of one of the dominant themes in both his popular and his scholarly writing: contingency in evolution. Like many other scientists, Gould frequently argues against the widespread idea that the course of evolution is somehow *necessary:* progressive, goal directed, always moving from the less to the more complex, and culminating in those marvels of reflective intelligence, ourselves. Again like many others, however, Gould (e.g., 1977:18) has no difficulty attributing these qualities of orderly goal direction to *development;* hence his use of the common metaphor of the computer program to explain developmental processes.[1]

The two words *evolution* and *development* have intertwined histories that reflect changing understandings of what we would now call the evolution of populations and the development of organisms. Much of this book is about the way the relationship between these two formational processes is conceptualized. I will not address the problems of mass extinction that are presented in *Wonderful Life,* but I will use Gould's discussion of the Burgess Shale creatures as a pretext for examining the way contingency is used to contrast evolutionary processes with developmental ones. I argue for a notion of development in which contingency is central and constitutive, not merely secondary alteration of more fundamental, "preprogrammed" forms.

Gould uses *contingency* in two major ways: as unpredictability and as a certain kind of causal dependency. It is useful for my purposes to distinguish the epistemological from the ontological sense of the word. I wish to examine the usual assumption that while chordates, or any other phylum, may be *evolutionarily* contingent, any particular chordate (or, for that matter, any organism) is hardly contingent *ontogenetically,* because it is brought into being by an internal plan. My discussion of developmental processes, which concludes with some thoughts on contingency in theorizing about developmental systems, also notes some possible connections to recent thinking in critical theory and the sociology of science. These connections startled me when I first confronted them. But perhaps they are not so surprising: Insofar as the modernist project consists of the attempt to separate nature from culture (Latour 1993), a serious rethinking of the two would seem very much to the point.

Contingency: Predictability and Process

Gould's argument in *Wonderful Life* is that the very existence of the phylum Chordata, and therefore of ourselves, is an evolutionary accident. He repeatedly uses the metaphor of the lottery and says that "this model strongly promotes the role of contingency, viewed primarily as unpredictability, in evolution" (1989a:308). It might seem that his primary concern is epistemological. This impression is strengthened when, in the same passage, he challenges his readers to contemplate the elaborate designs of these extinct creatures and to say what defect explains why they, and not others, vanished from the earth. But simple prediction (or retrodiction) is surely not Gould's only point. He is also making an argument about what kinds of *causal processes* are involved in evolution, "an unpredictable sequence of antecedent states, where any major change in any step of the sequence would have altered the final result. This final result is therefore dependent, or contingent, upon everything that came before— the unerasable and determining signature of history" (1989a:283–284).

Much could be said about this passage, but for now, we may note that Gould is joining the epistemological issue of predictability with the ontological one of the nature of evolutionary process, as chains of certain kinds of causal dependency. Gould certainly didn't invent the association between these particular meanings of *contingency* (Cahn 1967; Dray 1967).[2] Nor is the association always pernicious. As I noted above, however, the distinction between the predictability of processes and the nature of those processes will be crucial to my discussion of developmental dynamics.

Development involves the repeated arising and transformation of complexes of interacting processes and entities. These occur not because of preordained necessity, and not, obviously, by "mere chance," if this means an absence of regularity or causal relation. Rather, they come about through systems of contingencies whose organization may itself be contingent, and in any case, must be accounted for. Reliable, species-typical life courses can be seen as contingent in a number of ways (e.g., not absolutely necessary, causally dependent on factors that may in turn be uncertain). They may still be highly predictable, and thus *non*contingent, in another sense. Unlike evolution, development presents us with repeating cycles. Despite many kinds of variation, these can be so similar across generations that they offer themselves for comparison with the transmission of property by social institutions of inheritance. They thus invite confusion between the epistemological and ontological aspects of contingency. The reliability of these cycles tempts us to think that their products are somehow delivered in a package, ready-made and whole.

There are many other intergenerational bridges besides the chromosomes, but the genes are often endowed with extraordinary causal powers, sometimes being credited with the capacity to generate the next cycle virtually de novo. The sheer predictability of many aspects of development has been explained by invoking special causal processes, directed by molecular agencies that are immortal, omnipotent, omniscient (even immaterial; see Dawkins 1976; and Williams 1966). This involves a systematic privileging of insides over outsides and active controlling agents over passive materials: The environment is seen as supporting, modulating, and constraining development while primary formative power is reserved for the genes (Doyle 1997; Keller 1985; Oyama 2000).

An alternative is to explain such reliably repeated sequences of events by invoking heterogeneous, complexly interacting, and mutually constraining entities and processes, in which "control" is distributed and fluid rather than centralized and fixed (Gray 1992; Griffiths and Gray 1994; Johnston and Gottlieb 1990; Oyama 1982, 2000). The successive reconstruction of life cycles is possible not because constructing agents are protected from the outside world by the nuclear boundary, but precisely because what is inside the cell interacts with what is outside it. Once we acknowledge the many sources of intergenerational continuity and the repeating interactions that bring them together, we have no need for supermolecules to create the organism from scratch (Cohen 1979; Gottlieb 1976; L. Margulis 1981; West and King 1988).

Whether or not an element or variety of energy is a "resource" depends on its relation to the developing organism, which is in turn defined and constructed by its internal and external interactions. In this view, a gene is a resource among others rather than a directing intelligence that uses resources for its own ends. There is no centralized repository of "information" and causal potency that explains the repeated lives of organisms, no matter how much our notions of biological necessity may seem to require one. There are, however, many ways in which processes that have usually been considered mutually independent can be seen as actually impinging on each other. These relations are part of the developmental story.

If one adopts this way of approaching vital processes, many vexatious distinctions become not only unnecessary but unintelligible, while reasonable ones can be unambiguously characterized. One can compare, that is, features that vary within a species with those that are species-typical, or outcomes that seem difficult to perturb experimentally under a certain range of circumstances with those that vary in that same range, and so on. The point is that these are *different* questions, not different ways of approaching the same question (see chapters 3 and 7).

The notion of repeated cascades of contingencies, some more tightly constrained than others, has been central to work on developmental systems. Developmental influences interact over the life cycle to produce, maintain, and alter the organism and its changing worlds. I have sometimes called these influences, whether animate and inanimate, "interactants." Although this term was formulated independently of Bruno Latour's (1991) and Michel Callon's (1986) work on actants in "techno-economic networks," there are some striking conceptual similarities, which I believe to be nonaccidental, albeit unintended. To a certain extent their work stems from problems that resemble the ones that have engaged me, and that are historically and conceptually related to them.

In a developmental system, interactants and processes change over ontogenetic and phylogenetic time. Some are more reliable than others: The term *system* should not be taken as a guarantor of absolutely faithful replication, but rather as a marker of a complex, interacting network that *may* arrange its own relatively accurate repetition. *System* implies some degree of self-organization, in which "self" is not some privileged constituent or prime mover, but rather an entity-and-its-world, which world is extended and heterogeneous, with indeterminate and shifting boundaries.[3] Evolution, then, is change in these systems. For those who insist on the neo-Darwinian synthesis definition of evolution in terms of allele (variant gene form) frequencies, it is still possible to bring only these elements to the fore. To do so privileges the molecular level, however (in fact, only one aspect of it), and the gene-centered view has a variety of regrettable consequences, many of which are reviewed in these pages.

Developmental and Evolutionary Contingency

In speaking of unpredictable historical sequences, Stephen Gould (1989a: 278) says that "contingency precludes [their] repetition, even from an identical starting point." But developmental contingencies, and at least similar starting points, are what *allow* repetition, if and when it occurs. Causal dependence on uncertain conditions (one definition of contingency) needn't involve unique, unrepeatable, in-principle-unpredictable sequences. In the recurring life cycles discussed above, the conditions for various interactions are dependable to varying degrees. One needs to know when and how those conditions become more or less certain (How reliable are the formational processes?) and how crucial it is that precisely those conditions be present (How forgiving are they?). These questions are themselves related. Because formational robustness depends on other factors, one should not really speak of "canalized" or "buffered" characters as though this were somehow inherent in the character itself. Human limb formation is reliable under many, but certainly not all, gesta-

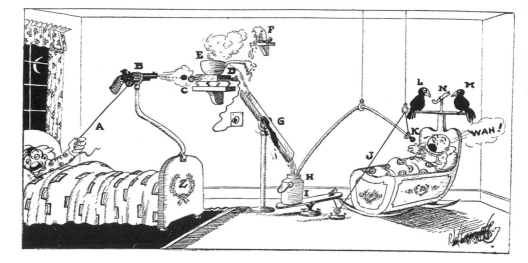

FIGURE 5. *Inventions of Professor Lucifer Butts: Anti-floor-walking para-phernalia, by Rube Goldberg (1932). Reprinted from Thomas Craven (Ed.), 1943,* Cartoon Cavalcade *(p. 235), New York: Simon and Schuster.*

tional circumstances; think of thalidomide and the limb deformities mentioned earlier. To continue the chain of conditionals, such circumstances have different effects on different people in different settings at different times. (Thalidomide in the eighth gestational month has different effects from those caused by the drug in the third month; its effects are likely to vary with other developmental factors, and there are surely individual differences in responsiveness.) Even bones are being constantly unmade and remade throughout life, and usage patterns are of great importance in their shaping.

Consider the example of an assembly of contingencies shown in figure 5. In an invention by cartoonist Rube Goldberg (1943b:235), the adult desirous of sleep must initiate the following sequence: "Pull string (A) which discharges pistol (B) and bullet (C) hits switch on electric stove (D), warming pot of milk (E). Vapor from milk melts candle (F) which drips on handle of pot causing it to upset and spill milk down trough (G) and into can (H). Weight bears down on lever (I) pulling string (J), which brings nursing nipple (K) within baby's reach. In the meantime baby's yelling has awakened two pet crows (L & M) and they discover rubber

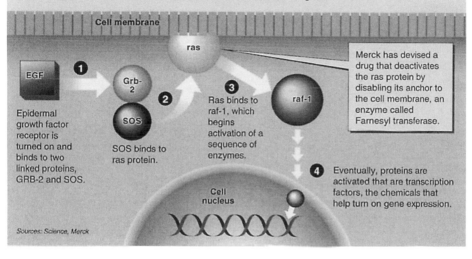

The Relay Race of Cell-Division Signals

The ras pathway is the principal circuitry through which a signal to divide reaches central headquarters in the cell's nucleus. New research illuminating cellular signals has made it possible to trace steps along the Rube Goldberg-style pathway. The series of molecular reactions begins when a stimulatory compound in the blood system tweaks a protein on the surface of a cell and ends with new cell growth.

Cell membrane

ras

EGF

1

Grb-2

2

SOS

Epidermal growth factor receptor is turned on and binds to two linked proteins, GRB-2 and SOS.

SOS binds to ras protein.

3
Ras binds to raf-1, which begins activation of a sequence of enzymes.

raf-1

Merck has devised a drug that deactivates the ras protein by disabling its anchor to the cell membrane, an enzyme called Farnesyl transferase.

4 Eventually, proteins are activated that are transcription factors, the chemicals that help turn on gene expression.

Cell nucleus

Sources: *Science, Merck*

FIGURE 6. *A segment of the* ras *(rat sarcoma) pathway thought to be involved in the development of cancers in a variety of species. Illustration by Baden Copeland. Reprinted from* The New York Times, *June 29, 1993, p. C10.*

worm (N) which they proceed to eat. Unable to masticate it, they pull it back and forth causing cradle to rock and put baby to sleep."

Figure 6 is another example of a chain of contingencies. Natalie Angier (1993:C1), of the *New York Times,* says of recent discoveries about the *ras* pathway in cell division (the *ras* gene, named for *rat* sarcoma, is believed to be implicated in many cancers):

Molecular biologists are now gazing upon their glittering prize, a fundamental revelation into how the body grows. The sight is astonishing to behold. It is epic.

It deserves a crash of cymbals, a roll of the tympanum—and a hearty guffaw. It turns out the much-exalted signaling pathway of the cell is a kind of molecular comedy, in which one protein hooks up to a second protein that then jointly push a button on an enzyme that pushes a button on another enzyme that makes this knob slide into that hole—all in all like something Wile E. Coyote might have pieced together from one of his Acme kits.

The joke here has nothing to do with unrepeatability. Angier continues: "Despite its improbability, its cartoonish complexity, the design works wondrously in overseeing cell growth, so well that it is shared by species as distantly related as yeast, worms, flies and humans" (1993:C10). Rather, the comedy seems to derive from the violation of our notions of simplicity, of logical necessity. One scientist involved in *ras* research asked plaintively: "Why is it all so complicated? Why do you need so many steps?" (Angier 1993:C10, quoting *ras* investigator Anthony Pawson). Biologist Sydney Brenner, introduced in chapter 1, made similar comments on his failure to find the developmental program for the tiny, much-studied roundworm *Caenorhabditis elegans.* Brenner complained that the worm's cell lines are "baroque" and that he could find no briefer way of describing what happens than simply giving an account of the entire sequence of events (in Lewin 1984:1327).[4]

The sequences in the *ras* pathway and in the ontogeny of Brenner's worm are "improbable" in that they would have been hard to predict before the fact. They have an arbitrary, uneconomical air that offends the sensibilities of those seeking the spare elegance of "law"; each event seems to occur not because of some transcendental necessity, but because the constituents just happen to be lying about. But the whole point, of course, is that they *do* lie about, enabled and constrained by a spatio-temporal organization that becomes possible only in a richly differentiated setting. Consider, for example, the enforced propinquity of diverse molecules in a structurally complex cell. These objects, relations, and reactions are capable of assembling, if temporarily, into systems of local necessities: contingently predictable congeries of dependencies and interdependencies, recurring cascades of contingencies.

The *ras* narrative takes place, for the most part, in the interior of individual cells, but these linked cascades occur on larger scales as well, and can include stable psychological and social processes, which in turn can feed back into physiological events. As long as "environmental" effects are understood as accidental, however, they are unlikely to be integrated into accounts of ontogenesis.

Goldberg's anti-floor-walking devices must be reset for their next use (though he also recommends that the adult keep earplugs handy just in case). Part of the functioning of many biological processes, however, is the entrainment of the next cycle. As noted above, this becomes possible precisely because the processes are *not* isolated, insulated, or autonomous, but are connected to others, by which they are influenced, and which they may in turn influence. Because these enabling/constraining systems do not stop at the skin, the developmental systems perspective requires a willingness to cross familiar boundaries in tracing such connections. This allows seasonal or other ecological regularities, for example, to be included in the developmental account. (See Griffiths and Gray 1994 for one way of individuating these complexes.) In a discussion of the history of life, Jack Cohen (1989:11) tells of chemical complexes that do "reset" themselves. These are self-feeding chemical reactions: "Each reaction affected some other process which led through a series of steps back to control of itself." The importance of such feedbacks, he notes, lay in the fact that they could "*keep the set-up going,* instead of hastening it to some kind of completion or exhaustion." These multiple dependencies ultimately make the metaphor of the linear chain inapt, though a scientist may excise part of the process to analyze as if it were an isolated chain running off autonomously against the background of the rest of the system. To do so, however, all of that background must be held constant (treated as *given* as well as kept from varying), just as Goldberg stabilizes by fiat, and is drolly mute about, many of the connections and contingencies that bring together, and hold together, his precarious assemblies.

Goldberg's contraptions also make ingenious use of just the kind of human-nonhuman-machine hybridization Latour (1991) and Callon (1986, 1991) mention in their respective analyses of networks. Notice that the desires and actions of Goldbergian crows must be assumed to be as stable as the melting point of paraffin. In another of Professor Butts's inventions a fan's components include not only a wheelbarrow and a doll, but also a live lovebird that nods every time it is asked whether it loves a mechanical bird, as well as a bear that, when "annoyed" at being kicked, "suspects" the doll dangling before it and eats it, thus triggering the next step in the astonishing concatenation (R. Goldberg 1943a:214).

Contingency in Theorizing Developmental Systems

We have seen that the epistemological question of predictability is not always distinguished from the ontological one of process in discussions of contingency. Yet reliably repeated assemblies can be noncontingent in the sense of being highly predictable while being thoroughly contingent in their dependence on complex, extended systems of interacting factors whose dynamic organization cannot be explained in terms of a single component or central agency. I conclude this discussion by suggesting that we extend the habit of remembering contingency, of saying "It depends . . . ," to the very activity of theorizing.

Characteristic of much of the work on developmental systems is an insistence on a sort of parity of reasoning. If, for example, some aspect of an organism is deemed to have a "biological base" because its variants are correlated with genetic variation in a particular population at a particular time, then one also ought to be willing to call something "environmentally based" if it is correlated with variations in the surround. (As some commentators on the nature-nurture opposition have pointed out, the same feature may thus be "wholly biological" and "wholly environmental," depending on conditions and the investigative choice of the researcher. See Johnston 1987.) Or, if some bit of matter is termed a "master molecule" because certain interesting things happen after it is activated (this is a loose inference about usage; the reasoning behind such terms is seldom explicit), then the same criteria should be used for all links in the sequence having similarly interesting sequelae—and one need hardly add that what is interesting, and to whom, is hardly fixed or uniform. (On master molecules, see Keller 1985:154, quoting David Nanney.) Obviously the point of such exercises in conceptual equity is not to distribute causal honorifics to more and more entities, but rather to question them by rendering them explicitly situation specific. By undermining the raison d'être of such terms, which is to elevate certain elements above others, the practice of relativization-by-contextualization should curb the impulse toward wildly extending the range of application.

One virtue of the parity-of-reasoning arguments, which in many ways resemble the "symmetry" rule described by Callon (1986) and by Barbara Herrnstein Smith (1991), for example, is that they direct attention to the often-unrecognized assumptions that inform the questions we ask and the ways we ask them. To emphasize the relativity of questions to the assumptions and purposes that lie behind them is not to claim that all entities, questions, or assumptions are equal or that it doesn't matter what one thinks because all ideas are equally valid.[5] On the contrary, precisely because it matters very much what questions we ask and why, and because the meaning of a question, as well as its possible answers, depends on what kind of world it takes for granted, it is important to consider the vision of the world implied by those questions. Even the boundaries of a developmental system are relative to the type of inquiry one is conducting, to the kind of story one wants to tell (van der Weele 1999:chap. 4). A social-psychological account may include elements that an evolutionary one ignores.

Consider a scar, perhaps one incurred in a duel. Paul Griffiths and Russell Gray (1994:286) would exclude scars from the developmental system because they would not contribute to their own reconstruction in the next generation (as a heart, for instance, would). But a scar carrying a certain symbolic or social weight might so contribute, at least in a short-term and general way, under certain social circumstances. A saber duel might give immediate resolution to a dispute, but it might also have a medium-term tendency to stabilize a person and family on the social landscape (with who knows what consequences for their reproductive success), and a longer-term tendency to legitimize and perpetuate the social arrangements themselves, including customary methods of resolving disputes, and thus the occurrence and the meaning of certain kinds of scars.

There are many degrees of transgenerational stability, so Griffiths and Gray's "evolutionary developmental systems," as they might be termed, would grade into the more inclusive sort that I have written about. Defined by the totality of causal interactions constituting a developmental trajectory, not just the ones that are regular across evolutionary time, these include features that play a role in a particular lifetime but do not recur in offspring. Either way, a system can include the sorts of factors that are typically of interest to a developmental psychologist, say, or a sociologist, as well as those that are more likely to occupy a physiologist or a geneticist. Many genetic mutations, for instance, would not meet Griffiths and Gray's criterion but could nevertheless have an important developmental impact on an organism. Such gradations are not only convenient, they are also theoretically crucial for undoing the distinction between cultural and biological evolution (chapters 1 and 9; Oyama 1994). This is a goal I share with Griffiths and Gray. Dueling scars may not recur across hundreds of generations, but circumcision, for example, can be transgenerationally stable for long periods and important in maintaining both cultural continuity and nonrandom reproduction. (Perhaps the ultimate such developmental "accident" is the navel, and it is a nice exercise to test the various criteria for nature and nurture on it.)

It is possible to demonstrate, then, that many ways of privileging the gene are unjustified. This can lead to including more kinds of phenomena in an initially restricted scheme, as when levels of selection or even replicators proliferate when a particular logic is applied to more and more cases.[6] But, as my earlier comments on master molecules suggested, it can also call into question the very language of self-replication, information, autonomous control—language that may become untenable when it is more fully contextualized. Consider again Angier's (1993) description of just some of the contingencies involved in the *ras* protein's workings, and her willingness to refer to just *this* molecule as "the mastermind of a signaling cascade." One could also, I suppose, call the rubber worm in Rube Goldberg's machine the mastermind of baby soothing.

Although Angier's account appeared in the *New York Times,* such language is common in the scientific literature as well. More to the point, it is tied to a way of thinking about biological processes that can be quite problematical. Such oppositions as autonomy versus connection, central control versus distributed control, and nucleus versus cytoplasm have rich metaphorical associations that are not scientifically or socially inert. The methodological, metaphorical, and wider social aspects of these matters are not really separable (Keller 1985, especially chap. 8, on the "pacemaker," 1992; Sapp 1987).

Latour (1987:71–72) has described the scientist as a spokesperson for that which is studied. One of the many reasons I have found it worthwhile to think and write in developmental systems terms is that it allows me to speak for the background—the mute, manipulated materials, the featureless surround. Sometimes, the peripheral is the political. Ultimately it may seem less appropriate to speak of entities' replicating themselves (thus marginalizing and instrumentalizing everything else) than to say that they may be assembled or constructed again and again as a result of processes in which they play a part but do not "control," except when one isolates a part of the process for analysis.

Seeing erstwhile prime movers as players in a game they control only according to a particular framing of a question and only by being controlled by other factors doesn't keep us from making distinctions, from studying processes or intervening in them. It may, as I suggest in chapter 8, make us less likely to underestimate the constitutive importance of these "other factors." We will not then be as surprised when it all turns out to be so complicated, or when unintended consequences and overlooked (but necessary after all) conditions (or people) force us to look again, for these are just what are pushed into the dim background by more monomaniacal stories (Star 1991).

By emphasizing the contingency of humans (introducing a distinctly unprepossessing Burgess chordate at the very end of his book), Gould denies predestination and draws a moral lesson about the importance of choice and action. But if theorizing about contingency is itself contingent in the ways I have suggested, we can turn Gould's moral around: It is equally important to recognize the "choices" we have *already* made, however unreflectively or tacitly. Indeed, it is essential to articulate them and own them, or even, once we have looked at them closely and related them to our other beliefs and concerns, to put them aside and make better ones. Taking some factor for granted or including it in a ceteris paribus clause doesn't mean that it plays no formative role or that it will always be there. This is something we realize with growing alarm as developmental, social, and ecological systems go awry, forcing closer attention to those "background" conditions that account for both the robustness and the vulnerability of developmental systems.

Part II

Looking at Ourselves

7 Essentialism, Women, and War:

Protesting Too Much, Protesting Too Little

In the 1970s and 1980s certain biological theorists and feminists converged on "essentialist" accounts of war that are strangely similar. At first glance it seems an unlikely development, given the frequency with which feminists and antifeminists have lined up on opposite sides of the nature-nurture rift. At second glance, though, perhaps the alliance is not so surprising after all. We seem to be in the midst of a pendulum swing "back to nature." This movement, in turn, is probably part of a more general trend in the United States toward conservatism and a certain brand of romanticism, although the issue is complex. Apart from the current emphasis on so-called traditional values, though, the convergence on "biological" views reflects some very pervasive beliefs that are as evident in environmental determinist approaches as they are in biological ones.

By "essentialist," I mean an assumption that human beings have an underlying universal nature that is more fundamental than any variations that may exist among us, and that is in some sense always present—perhaps as a "propensity"—even when it is not actually discernible. People frequently define this preexisting nature in biological terms and believe that it will tend to express itself even though it might be somewhat modified by learning, and thus partially obscured by a cultural veneer.[1]

When I say that social determinist and biological determinist approaches share many assumptions, I mean that even when they have disagreed on the proper "balance" of cultural and biological explanation, they have treated the general project of finding the correct proportions of environmental and genetic influence as a valid one. They have also tended to agree that the possibility of change was somehow illuminated by their disputes.

In a revealing metaphor, sociobiologist David Barash (1979:12) compares the relationship of nature and nurture to two people wrestling. As they tumble about, he says, their limbs entwine so that it is hard to tell which is which. Yet, however entangled they may become, the combatants do not merge; they are separate persons in competition, and our imperfect powers of observation do not change that fact. We will return to this problem of separateness later. Let's look first, however, at some examples of essentialist accounts of women and war. The first several come from scholars who have offered us their biological views of human behavior and society; the last two come from a 1982 collection of feminist writings.

The Argument

Lionel Tiger and Robin Fox (1971:212–213) say that war "is not a human action but a male action; war is not a human problem but a male problem." If nuclear weapons could be curbed for a year and women could be put into "all the menial and mighty military posts in the world," they say, there would be no war. Tiger and Fox immediately concede that this proposition is a totally unrealistic fantasy, because the human "biogrammar" (a term they use more or less the way others use "genetic program") ensures that such a thing could never happen. Men, they say, have evolved as hunters who band into groups and turn their aggressiveness outward against common enemies or prey. Political structures in modern societies are formed on this primeval hunting model, and men naturally dominate these structures as well. Tiger speculates that women, who he says do not bond and cooperate as effectively as men, could be given positions in government by special mandate. He feels, however, that it might be unwise to expect that even this effort could effect much change (1970:270–272). Presumably, attempts to subvert natural tendencies are not likely to succeed.

Barash (1979:187–189) accepts the idea that males bond and exclude women from political power, although he emphasizes the grounding of male aggression in competition for reproductive opportunities and reproductively relevant resources (p. 174). He argues that women are allowed to hold political power only if they are in some sense "desexed" by age, physical unattractiveness, or both. Otherwise, men refuse to recognize a woman's authority, even when she manages to gain admittance to the male "club" (pp. 189–190).

Another commentator on matters biological, Melvin Konner (1982: xviii, 126, 420), offers the idea that women, because they are less aggressive than men, should be placed in authority in order to "buffer" or "dampen" violent conflict between nations. He seems to reject hunting bands as the evolutionary explanation for human aggression, although he does cite Tiger's work and allows that "something happens when men get together in groups; it is not well understood, but it is natural, and it is altogether not very nice" (pp. 203–206). Like Barash, Konner (chap. 12) is more impressed with the notion that male aggression is explained by competition for the reproductive resources provided by females: eggs and parental care.

The suggestion is thus made that women might be more peaceful than men in positions of power, but it is immediately and quite effectively undone by these theorists' other assumptions about natural differences between women and men, and about the social consequences of these differences. It is a rather neat irony, that the qualities that might save the world are kept out of the public sphere by the very same biological order that produces them. Konner (1982:126) does not say that women must *always* be excluded from power, although he asserts that it is pointless to use violent female rulers of the past as models for the future, because they "have invariably been embedded in and bound by an almost totally masculine power structure, and have gotten where they were by being unrepresentative of their gender." He does not say how to implement his recommendation that "average" women be allowed to control the world's arsenals. (Perhaps it would be easier to recruit abnormally peaceful men.)

The argument that women's exclusion from politics is inevitable also appears, of course, in past and present antifeminist writings on the necessity of patriarchy.[2] Partly because biological arguments have often been associated with reactionary politics, it is now common for theorists to declare their liberal values, deny that biological treatments are necessarily either deterministic or conservative, and emphasize that biological explanation is not the same as moral approval. Then the theorists typically call on us to know our natures in order to transcend them. Barash (1979:198), for example, asks whether we can use our understanding to overrule the biological "whisperings from within." At the same time, these writers often warn *against* trying to challenge the boundaries and constraints our genes set for us. Charles Lumsden and E. O. Wilson (1981:359–360) warn that trying to escape these constraints risks the "very essence of humanness." They advise us to learn what the limits are, and to set our goals within them, while Barash declares that denying natural sex differences is "likely to generate discontent" (1979:116).

The relationship between politics and science is a complex one, and although I think it can be argued that at any particular time some scientific approaches tend to be associated with or to imply certain attitudes toward the moral and political worlds, there is no direct tie between reactionary values and an interest in, for example, evolutionary analysis. Biologists become quite as annoyed as anyone else at having their politics misrepresented, and to assume that someone who emphasizes biological bases of human behavior is automatically a "crypto-Nazi" is to engage in just the sort of reductive thinking I am criticizing. A crucial link between one's scientific and political views is one's conception of will and possibility. Such conceptions are rarely made explicit, and yet these are just the sorts of hidden assumptions that structure the arguments and invite the conclusions.

Turning to the feminist literature, we find examples of essentialism in Pam McAllister's (1982) collection, *Reweaving the Web of Life*. In "The Prevalence of the Natural Law within Women," Connie Salamone describes women's roles in protecting both the young and the natural law that governs the world. This role, if it is not subverted by male values, endows females with a special affinity with other animals and tends to give rise to concern over animal rights and to vegetarianism. Salamone (1982:365–366) contrasts the female "aesthetic of untampered biological law" with "the artificial aesthetic of male science." In another paper in the McAllister collection, Barbara Zanotti (1982:17) invokes Mary Daly's concept of women's biophilia, or love for life, and goes on to describe the history of patriarchy as the history of war. She asserts that, in making war, patriarchy does not really attack the opposing military force; rather, it attacks *women,* who represent life. Soldiers, she suggests, are encouraged to identify military aggression with sexual aggression, so that "the language of war is the language of gynocide."

The argument under discussion, then, is that women are inherently less aggressive than men, that war is caused by male aggression, and that women are thus somehow more capable than men of bringing about peace—or, at least, would be less destructive than men if they were in power. It is a skeletal argument, of course, abstracted from very different sorts of writings. The transition from individual to international conflict in such works is not necessarily direct. For Tiger and Fox (1971), Barash (1979), and Zanotti (1982), for example, war is specifically the aggression of male-bonded men in *groups;* and Tiger (1970:219) distinguishes between individual and group violence. Furthermore, the connection between aggression and peacemaking is not clear. Especially in the work of the male scientists cited above, it is females' *lack* of aggressiveness, rather than any positive quality, that is emphasized. Even in a world in which women have traditionally been defined by their deficits, I'm not sure that peacemaking and peacekeeping should be seen as merely passive (to invoke another loaded dichotomy) results of low levels of aggression. Radical feminists are more likely to implicate the positive female qualities of nurturance, sensitivity to connections, and peacefulness.[3]

Lumping: How to Ignore Important Distinctions

The arguments about the sexes and aggression discussed above all require that aggression be unitary. They depend on, and encourage, certain kinds of illegitimate "lumping." One sort of lumping is definitional: All sorts of behavior, feelings, intentions, and effects of actions are grouped together as aggressive. Tiger's (1970:203) definition, for example, is so broad that it embraces all "effective action which is part of a process of mastery of the environment." Violence is one outcome of such aggressive activity, but not the only one. That women do not bond in aggressive groups implies, then, that they are less capable of "effective action" in the service of "mastery of the environment"—a sweeping statement indeed, and one that bodes ill for any political *action* on the part of women.

Another habit that allows us to treat aggression as a uniform quantity is cross-species lumping, equating very different phenomena, so that mounting and fighting in rodents, for example, territoriality in fish or birds, and hunting, murder, and political competition in humans are all "aggressive" behaviors. Then there is the lumping of levels of analysis, in which the activity of nations and institutions is reductively collapsed to the level of individuals, or even of hormones or genes. Finally, there is developmental lumping, in which activity levels in newborn babies, rough-and-tumble play in young children, fighting or delinquency in teenagers, and decisions of national leaders in wars are viewed as developmentally continuous, or somehow "the same." Sex differences in these behaviors are then seen as manifestations of basic sex differences in aggressiveness.[4]

Much of this lumping depends on very common modern-day versions of preformationism and essentialism. Today we think of preformationism as an archaic relic of outmoded thought, and we snicker at the absurd idea that there are little people curled up in sperm or egg cells. But replacing curled-up people with curled-up blueprints or programs for people is not so different. That is, there is not much conceptual distance between *aggression* in the genes, on the one hand, and *coded instructions for aggression* in the genes, on the other. What is central to preformationist thought is not the literal presence of fully formed creatures in germ cells, but rather a way of thinking about development—development as *revelation* of preformed essence rather than as contingent series of constructive interactions, transformations, and emergences. It is a way of thinking that makes real development irrelevant because the basic "information," or form, is there from the beginning, a legacy from our ancestors.

Nor is the basic reasoning much changed by the less deterministic-sounding language of biological predispositions, propensities, or limits to flexibility. As we have seen, the assumptions underlying these apparently moderate formulations are not substantially different from more dichotomous ones. Similarly, saying that of course nature combines with, or interacts with, nurture suggests a continued reliance on a biological nature defined before development begins and merely modulated or deflected by environmental nurture. One problem is that, in more moderate-sounding accounts, the genes still define the boundaries within which action is possible, and they still constitute the ultimate source of control. Barash (1979:200), for instance, likens humans to horses being ridden by genetic riders who give us considerable freedom but remain in firm command. How are those who hold this view to conceptualize a "we" that is pitted against our genes in a struggle for control over "our" behavior? Another puzzle, at least for feminists who embrace the argument for inherent male aggression and dominance, is how to mobilize for change in a world populated by inherently aggressive and dominant males.

What Is the Point?

It is important to be clear about what I am *not* doing when I criticize the nature-nurture opposition or when I decry the lumping of different definitions, species, levels of analysis, and developmental phenomena. I am not saying that aggression, however it is defined, is unimportant. I am not denying that nations are composed of individuals, and that individuals are composed of cells, chemicals, and so on. Indeed, understanding these parts might help us to understand the wholes of which our world is composed. I am not, therefore, rejecting research on individuals, hormones, neurons, and genes, including those of other species. I am not making an environmental determinist argument that biology is irrelevant, that genes don't count, and that everything about behavior can be changed. These last three, by the way, are not the same argument, and one of our problems is that we tend to lump them at the same time that we lump biological arguments. I am not even denying certain constancies or similarities among individuals within and across societies.

What I *am* saying is that analysis should be conducted in the interests of the eventual synthesis of a complex, multilevel reality (just as temporary lumping—of diverse essentialist treatments of aggression, for example— can serve the elaboration of a more complex argument). The levels I have in mind are not like onion skins that can be stripped away to reveal a more basic reality. After all, when one takes away enough of an onion's layers, there is nothing left to reveal. Rather, they are levels of analysis whose interrelationships must be *discovered,* not *assumed.* We will never understand the role of genes and hormones in individual lives or of individuals in society unless we move beyond traditional oppositions. We will never gain insight into the possibilities of different developmental pathways if we assume them to be fixed. This is the point at which the environmental and biological determinists, as well as the more moderate "in-betweenists," are unwitting allies: They usually agree roughly on what it would mean for something to be biological or cultural, even as they argue about relative contributions of genes and learning.

If we want to use scientific analysis to answer questions, we must know what questions we are asking, or we'll never know what evidence could help us answer them. And if we want to fight the good fight, we must know what the enemy is, or we will waste precious time and energy. Note that I say *what* the enemy is, not *who;* I am concerned with ways of thinking, not people.

Questions, Concerns, and Answers

We should not ask if something is "biological" and expect a single answer. We might be asking about the chemical processes associated with some behavior, for instance; this is a matter of the level of analysis, and such questions can be asked about any behavior, learned or unlearned, common or uncommon, fixed or labile. We might be asking about development: Does a given behavior seem to be learned? Is it present at birth? (These two are not the same thing, since learning can be prenatal, and not all postnatal changes involve learning.) We might be asking about evolutionary history, which in turn resolves into other questions: Is the behavior present in phylogenetic relatives? When did it appear in our own evolutionary line, and why? We might be asking what role, if any, a character now plays in enhancing survival and reproduction. We might be asking whether variation in a character is heritable in a given population: Are differences in the character correlated with genetic differences?

None of these questions has any automatic bearing on any other, and lumping them together as "genetic" or "biological" simply muddles matters. Often, however, a person who asks whether some trait is biological is not interested in these particular questions at all, but has something else in mind: the *inevitability* of a trait; or its unchangeability in the individuals evincing it; or its goodness, justifiability, or naturalness; or perhaps the consequences of trying to change or prevent it. Will it come bursting out as soon as we drop our guard? Will intervention do more harm than good? Scientists frequently share these concerns, as well as the assumptions that link them to biology. But worries about inevitability are really about *possible* developmental pathways, not about past or present ones. (Even wondering whether a present state of affairs is immutable implies wanting to know what would happen if . . . , and wondering whether it was inevitable implies wanting to know what would have happened if . . .) When Barash (1979:88) speculates that male parenting in humans is "not nearly as innate as modern sexual egalitarians" think, he seems to be commenting on the probability of reaching certain personal and political goals, and he seems to believe that "innateness" (a concept he never defines satisfactorily) has something to do with the difficulties that he thinks "sexual egalitarians" will encounter.

Inevitability is not predictable from observations at the morphological or biochemical levels of analysis. It is not predictable from the role played by learning in the development of a behavior, or from its time of appearance. It is not predictable from a phylogenetic history, a pattern of heritability in some population, prevalence in certain environments, or even from universality in a species. That is, none of these traditional scientific biological questions is relevant to the concerns that most often motivate the questions. To ask biology to tell us what is *desirable,* furthermore, is to ask science to do our moral work for us.

We must decide what kind of world we want, and why. We won't necessarily succeed in bringing it about, but we shouldn't be deterred prematurely from trying because of biological evidence of whatever variety. That is, we shouldn't simply leap to the conclusion that matters are hopeless, either because we believe the biological, in any of its senses, is fixed, or because we believe it is dangerous to tamper with what is "natural." Similarly, and just as important, we shouldn't be complacent about natural features we might value: Virtues that are thought basically feminine in this world, for instance, won't necessarily persist in the one that's coming. There is a tendency to view the biological as static, but it is, in fact, historical at all levels. The habit of asking whether some feature of our world is the result of biology or history is thus deeply odd. When people ask about biology, though, their concerns tend to be mythological, not historical. Here I do not mean *myth* as wrong, or "bad science," (though it might be), but as a way of thinking that hankers after ultimate truth, eternal necessity, and legitimacy.

Lionel Tiger, chronicler of male bonding and aggression, refers to *Lord of the Flies,* William Golding's widely read story of a group of English schoolboys marooned on an island who rapidly degenerate into a horde of savages. Apparently, Golding said that he wanted to construct a myth, a tale that would give the key to the whole of life and experience (cited in Tiger 1970:207). The feminist Zanotti, too, accepts the centrality of male bonding to individual and social life in her claim that, in making war, men are eternally attacking and destroying women (1982:17). Both theorists invoke unchanging essence to explain gender and the relations between men and women, and thus the world. But it is a static world in which ancient tragedies are played out again and again, according to primal necessity. It is not a historical world in which necessity and nature arise by process and then give way to other necessities and natures.

Playing the Game

It should be clear by now how I feel about several common strategies for dealing with biological arguments. When someone says, "It's biological," we sometimes reply, "No, it's not, it's *cultural,*" when instead we should be asking why the cultural and the biological are treated as alternatives in the first place, and just what we (and they) really mean by either explanation: Not all environmental influences are cultural ones, for instance. I call this the Protesting Too Much Syndrome, because we are often afraid that the trait in question *is* "biological" in one or more of the mistaken senses described above. Similarly, when someone says that women are innately inferior, we counter, "No, we're not, *you* are," rather than rejecting the assumption of essential nature that allows pronouncements of this sort. This second strategy, the Protesting Too Little Syndrome, entails agreeing that differences are biological, but reversing the evaluative polarity. Male nature is bad; female nature is good. While it offers the momentary satisfaction of turning the tables, it is based on all the ideas about nature that get us into so much difficulty. This is too great a price to pay for Mother Nature's favor. The solution is not to protest precisely the correct amount, or to find the degree of biological determination that is *just right,* like Goldilocks trying to find comfort in a house that is not her own. Rather, it is to protest *a whole lot* about the very rules of the discourse.

I am not saying that we ought to throw out everything and start from the ground up. We couldn't do it even if we wanted to. But when there are ample grounds for doubting the validity of a conceptual framework or a set of issues, as is the case with the nature-nurture complex, we do ourselves no favor by blindly accepting the terms of the game. Some of our gravest problems come, after all, from letting others set terms for us. The burden of clarification certainly does not rest entirely with women, but if we shirk our part, how can we do justice to the struggle?

Instead of pitting one mythical account against another, instead of searching for a morally or emotionally resonant evolutionary past to explain the present, and then projecting it into the future, we must focus on real historical processes whose courses are not foreseeable on the basis of any account of nature as manifested in hunter-gatherers, baboons or chimps, hormones, brain centers, or DNA strands. I speak here of individual developmental history, as well as historical change on the societal level, for it is within these processes that nature and possibility are defined.

Are Women Less Aggressive and Hence Less Warlike?

I could say more about aggression and about women, and maybe even a little about war. Perhaps the reason I haven't done so up to this point is that the essentialist theories I have been discussing don't say much about these topics either. Instead, I have focused on the ways we *think* and *talk* about these topics. War is about politics, diplomacy, economics. It is about historical continuity and change in relations among people, not about brain centers, testosterone levels, or rough-and-tumble play. War is like a fight between individuals only by analogy, just as certain encounters between groups of ants are wars only by analogy. Perhaps it is significant that the book sociobiologist Barash coauthored on preventing nuclear war (Barash and Lipton 1982) contains nothing about differing capacities and contributions of males and females, but instead lists very pragmatic suggestions for effective action. I would never claim that women have no role in national and international politics, but neither can I make sense of the notion that we ought to be somehow inserted into public life because of some mythic direct line to life, peace, and love. Men are not a plague, and women are not a cure.

8 The Conceptualization of Nature:

Nature as Design

The controversies I find most interesting are the ones that are notable for the questions they raise rather than the solutions they produce. One can argue about whether computers have minds or not, for instance, but the more intriguing question is, What does it mean to *have* a mind?—and indeed, Is a mind something one can "have"? Or one can wonder whether or not apes have language, but the real challenge is to say precisely what language is. Thus far we have often been concerned with the endless debate over whether or not some trait is "in" the genes: We have considered just what the phrase (or its opposite) could possibly mean, and what it shows about the way we think about the nature of organisms and their development. Examination of these questions reveals some unspoken assumptions. To begin with, they have spatial implications. Minds and language are thought to be things we either possess or don't; they have a location, typically in our brains. "Genetic" traits (or programs for them) are in the DNA. Behind these assumptions is the image of isolated individuals whose properties (possessions!) can be enumerated without paying much attention to their activities or surroundings.

To use biology as a basis for design—to ask it to guide our activities, in devising a more reasonable agriculture, say, or ecologically safer methods for processing waste, is to turn outward, not inward, but a similar situation arises. On the one hand we can wonder just which designs are biologically natural and which are not. Or we can ask whether nature has a design "there" for us to imitate. To turn Gertrude Stein's dismissive comment about Oakland, California, into a query, is there a *there* there? Does Nature have a nature, an essence, independent of us but knowable by us?

Small-*n*-nature, Big-*N*-Nature

My reflections in this chapter will be a departure from the rest of the book in two ways. First, my investigations of the relations between the internal and external domains have concentrated on the concept of internal biological nature. I once called this "small-*n*-nature," as in "human nature." Here, though, I turn my attention to Nature "out there": *Nature* with a capital *N* (these terms are from Oyama 1987). Second, instead of indicating the dangers of multiple meanings, as I did in chapter 7, I will celebrate ambiguity. My emphasis on constructive interaction remains. This time, however, it is applied not to the ontogeny of organismic "designs," but to the designs we see in the external world.

By insisting on construction, I don't mean that we simply project our internal designs onto a passive, chaotic Nature. This would be to accept the objectivist-subjectivist split criticized by Humberto Maturana and Francisco Varela (1987), by certain feminist theorists (Harding and O'Barr 1987, for instance), by Lakoff and Johnson (1980), and by others who are dissatisfied with what they consider to be seventeenth-century notions of objectivity. Rather, external *Nature*, like internal *nature*, is co-constructed over time, through intimate engagement with the world. Nature is multiple but not arbitrary, and design is *brought forth* with the world in that fusion of knowledge and action also described by Maturana and Varela (1987). Just as I have denied the validity of traditional notions of immutable biological nature inside us, written in our genes at the moment of our conception and existing independently of our developmental interactions, I now question the idea that big-*N*-Nature has a unitary, eternal nature that is independent of our lives in the world. In fact, the kinds of interactions in which we participate influence the Nature we design. (The ambiguity in the preceding sentence is intentional: I mean to refer both to our conceptualizations of Nature and to the kinds of concrete changes we bring about in the world.)

I am going to make some comments that may seem subversive to the notion of "biology as a basis for design," but ultimately I don't think they are. Along the way, I will discuss concepts of design and nature and the virtues of multiple perspectives in elaborating these concepts. I will touch on biology as a way of knowing, the relationship between knowers and knowledge, and the necessity of taking responsibility for our contributions to the knowledge (and therefore the Nature) we construct. This may seem rather a lot, but I wish only to sketch a set of issues, not to provide an exhaustive treatment, so my comments on each topic will be brief.

DESIGN IN NATURE

The word *design* is multiply ambiguous in English. It can be a noun or a verb, and it has multiple meanings in both roles. In creative activity, the word refers to the originating intention: One conceives a design, a plan or scheme. It also refers to the finished product: A completed design may be placed on display in a museum. Finally, it is the act of creation that fulfills a design: One designs (makes) a table.

In working from a preexisting model, on the other hand, as is implied by the idea of using biology as a basis for design, *design* refers to the preexisting external model: Replicas can be made of an original design. The finished product may also be called a design, although we usually take pains to distinguish the original from the copy. The activity itself, however, is not called *designing,* because the design is already there. Instead, we speak of *imitating,* of *copying* (see chapter 2).

The vocabulary reveals our assumptions. Subjects are treated as radically separable from objects. Design, or form, originates either inside or outside the subject-agent. A strict distinction is made between active creation and passive imitation, between originating a design and serving as a conduit through which it passes (chapter 3; see also Maturana and Varela 1987).

But we are not committed to these assumptions in order to consider nature and design. We can admit our interactive role in the definition of problems, in the choice and conceptualization of model, in the mode of investigation, in the construction of knowledge itself. I would even playfully suggest that we should *increase* the ambiguity of *design* by including *imitation* or *study,* or even *perception,* in the semantic complex that already embraces the creation of a design, the design created, and the design that guides our work. Who knows? This little exercise in creative muddling might even dignify imitation.[1] In any case, these reflections on design should help to undermine some of the traditional assumptions about knowledge and activity that are now being strongly challenged from many directions.

MULTIPLE PERSPECTIVES

Two incidents prompted my own thinking about these issues. One was a comment by Mary Catherine Bateson about the much-used metaphor of the earth as mother. She offered some alternatives—the earth and ourselves as co-parents, for example, or the earth as child—and made the gentle suggestion that we needn't insist on only one. There might be some virtue, that is, in a kind of tentative flexibility in our thinking about this relationship—or rather, these relationships.

The second prompt was a story in the *New Yorker* by Ursula Le Guin (1987). Entitled "Half Past Four," it is not so much a story told from several points of view as it is a story that is itself progressively transformed. In a series of scenes, people and events shift. Ann is first presented in an encounter with her father, Stephen, and with Toddie, the retarded son of her father's new wife, Ella. In the next scene Ann is Todd's sister, and the two siblings discuss their absent father, who has started a family with another woman. Then Ann is Ella's daughter, meeting Ella's suitor, Stephen; and so on. Identities, histories, and even sexual orientation are permuted and transmuted while the many-dimensioned space is bound together by recurrent themes and images.

Reading the story for the first time, I found myself backtracking impatiently in the careless reader's attempt to recover what an initial scan had missed. I fussed: Are Toddie and Ann siblings or half-siblings? Who is this Ella, anyway? It was only after several of these truth-gathering forays that I realized that getting the facts straight *once and for all* was not the point. Rather, the concatenated scenes could be read for their richly proliferating relations, including the relations among those relations.

These two experiences led me to reflect on the value of multiple perspectives—not as several lines of sight converging on the same object, in which the goal is a single, more accurate view, but rather as paths to richness, to curiosity, to a sense of possibilities and myriad connections, divergences, and discontinuities. Anthropologists Gregory and Mary Catherine Bateson speak of the diversity of reference, of point of view, in ethnographic knowing (1987:185).[2] (Perhaps it is only coincidental that

Le Guin is also the child of anthropologists.) It can be a corrective to the exclusionary absolutism of much of traditional science and, incidentally, much of traditional religion. Both of the latter, after all, are involved in legitimating only certain kinds of knowledge, in sanctioning only certain kinds of knowers and ways of knowing.

In scientific knowledge, the knower paradoxically disappears. The ideal of scientific objectivity is based, in fact, on the interchangeability of knowers. Sandra Harding (1986) has written about Baconian objectivity as universal subjectivity — knowledge by anybody — and about the corresponding scientific division of the world into the real (public, shared) and the unreal (merely private). An influential contemporary version of this division is the distinction between the biological (objective, real, physical, and basic) and the merely psychological or cultural (subjective, less real, evanescent, and arbitrary), and it is biologists' role as arbiters of the biologically real that lends them special authority in today's world.

The disappearing knower supports the myth of the autonomy and separateness of the world. That that separation may involve a degree of discomfort and insecurity is a possibility explored by Susan Bordo's account (1987) of the rise of Cartesian rationalism. She compares this historical development to the drama of separation in psychological development and notes that one way of reducing the pain of separation is aggressively to pursue separation; the pain is then "experienced as autonomy rather than helplessness in the face of the discontinuity between self and mother" (p. 259). I would read this, by the way, as Bordo herself does, not as a developmental psychopathology of science, but as "a hermeneutic aid" enabling us "to recognize the thoroughly historical character of precisely those categories of self and innerness that describe the modern sense of relatedness to the world" (p. 253).

The Conceptualization of Nature

BIOLOGISTS AS KNOWERS

Biologists are heirs to this tradition of knowledge, and they frequently exhibit the traditional reluctance to recognize their own contributions to it. Such recognition threatens, they fear, to make knowledge "merely" subjective, a solipsistic projection on the world. The scientific method, after all, is about eliminating subjectivity, about protecting pure factuality from "bias." These fears about the contamination of truth assume just what I wish to question: the separation of the knower from the known, the opposition of active creation to passive reception, and the conception of pattern and design as things with independent existences. If we take the virtues of multiple perspectives seriously, we must own up to our concerns and sensitivities, embracing a more interactive view of knowledge and action. Despite our fears, I believe that neither we nor the world will disappear.

I am not just complaining about "bad science," a science infected by extraneous interests and points of view. This implies that a pure, depersonalized science can exist. Rather, this story making from a particular point of view is what science is—not only "bad science" but also what Helen Longino and Ruth Doell (1983/1987) call "science as usual." [3]

INFORMATION

Information is the modern, technological incarnation of the notion of design; it is thus central to the issues at hand. Information is usually conceptualized as a kind of stuff that can be found both inside and outside of us. The information in us, revealed in "biological bases," defines the instinctive core of our being. It is a prescribed inner reality: small-n-nature, typically thought to be encoded as genetic information that is "translated" into bodies and minds. Information "out there," on the other hand, comes from the external world. It is supposed to be acquired through perception, and it then appears in the mind as representational knowledge. Somehow information moves from big-N-Nature into our heads.

But information is not some mysterious stuff, capable of being transmitted from one place to another, translated, accumulated, and stored; rather, it grows out of kinds of relations. For Gregory Bateson (1972), information is a difference that makes a difference. This invites questions: a difference in what (What are you paying attention to?), about what (What matters?), for whom (Who is asking, who is affected?). Asking these questions leads us to focus on the knower, a knower who always has a particular history, social location, and point of view.

Ironically, the science that was based on the democratization of knowledge has produced a technocratic elite, so that knowledge-by-anybody is actually knowledge by a very few. The habitual disembodying of scientific knowledge, however, tends to obscure the specialness, the exclusiveness, of this class of knowers. This mystification-by-depersonalization is exemplified by the frequent use of the passive voice in scientific writing.

If we acknowledge that our interests, perspectives, cares, and worries are part of the complex in which information is generated, then the knowing subject can't disappear as easily as it does in conventional science. It should become harder to mystify knowledge and action, and easier to detect the politics of knowledge: the subtle power involved in defining problems and evidence, in legitimizing knowers and knowledge. Women and non-Westerners are often outside the inner circle of scientific knowing, for instance, and farmers' knowledge has been largely excluded from scientific agriculture (Jackson 1987).[4] In addition to highlighting the usually invisible power relationships that partially structure our knowledge, recognition of our active role in knowing may lessen the chances that we will be reductively stuck in one perspective, or at one level of analysis.

Biology as a Basis for Design

What kind of biology? What kind of basis? What kind of design? If we mean a biology that is fixated on a small-n-nature inscribed in the secret code of base pairs, if we believe that Nature designs organisms by giving them genes for things, and that we can imitate her by inserting those genes into other organisms ("reprogramming" them) to make them produce the same things, then we already have a certain kind of biology as a certain kind of basis for a certain kind of design. Just as surely, our concepts of design and of ourselves have served as bases for that kind of biology. (See Haraway 1985b for a spirited commentary on the modern constitution of organisms as "cyborgs," or cybernetic organisms.) If we mean a biology that is fixated on a single vision of big-N-Nature, perhaps as ruthlessly competitive or as inherently cooperative, as a perpetual race or as harmonious equilibrium, and if we believe we can order our own lives and activities by this vision of Nature, then we already have this kind of biology as well; it is a basis for, and a reflection of, our ideas of (or aspirations for) our own design.

Can we envision another kind of biology, one less tied to a search for the one timeless truth that will structure our lives, a biology that recognizes our own part in our constructions of internal and external natures, and appreciates particular perspectives as vehicles for empathy, investigation, and change? If a metaphor is something through which we can think (G. Bateson and Bateson 1987:chap. 17), and perhaps through which we can create/discover ourselves, can we find different metaphors for our world, and thus construct different ways of being in it?

TAKING RESPONSIBILITY FOR NATURE

We must take responsibility for the Nature (and the biology) we construct. We do not, however, manufacture either our own natures or Nature out there as detached, Godlike subjects. Our responsibility is not the responsibility of unmoved movers, absolute originators bringing order to chaos. Rather, the construction is mutual; it occurs through intimate interaction. By the same token, we do not simply record facts about external Nature, any more than we are simply manifestations of an internal nature encoded in some genetic text. "Information," that is, is not independent of us, and because this is so, we cannot disclaim a kind of ownership. Our cognitive and ethical responsibilities are based on our *response-ability,* our capacity to know and to do, our active involvement in knowledge and reflection.

This is a productive ambiguity of subjects and objects, of multiple perspectives, of ourselves as nature's designers and as nature's designs, as designers of our designs of (and on) nature, of our own natures as products of our lives in nature. If Nature is our technical or moral teacher, it is not as a radically separate, independent source of information-stuff, but rather as a source of differences, a world we interrogate with particular questions and particular concerns in mind, and whose responses we interpret in the light of our own history with it.

I respect and value much of the work on alternative biologies, but I also think we ought to be cautious about claiming one true account of Nature's nature (although I have in the past come close to doing this myself). I am suspicious of quests for single truths; there is a strong tendency to take what is pre-scribed ("already written," whether in the genes or in the language of the earth and streams, and therefore in no way dependent on us) as *prescription,* that is, as a received formula for action. We should re-sist this temptation to seek "the natural" as an authoritative guide to our lives. It is, after all, the temptation to deny our presence in every truth we see. Our presence in our own knowledge, however, is not *contamination,* as some may fear, but the very *condition* for the generation of that knowl-edge. Biology is one activity among many, and design is the pattern that emerges with that activity. Finally, this includes even our conceptions of Nature's own design.

Donna Haraway writes of science as fetish: "an object human beings make only to forget their role in creating it, no longer responsive to the dialectical interplay of human beings with the surrounding world" (1987:219). Elsewhere, she calls on us to forsake our search for a total-izing unity, to give up our "dreams for a perfectly true language, of per-fectly faithful naming of experience" (1985b:92). In place of the quest for a lost innocence, an original wholeness to make our lives intelligible and to ground our politics, she advocates a politics of coalition. Such a coalition would be based not on a common language, but on a "powerful infidel heteroglossia" (1985b:101).

Much of what I have just said could be seen as a denial of the desir-ability, or even the possibility, of taking biology as a basis for design. In fact, though, once our terms are reworked in a way that fits the enterprise, just the opposite is true. It is "biology" as timeless truth that I reject, not biology as a human activity entered into with responsible awareness. The first defines reality once and for all; the second is closely connected to the persons and life circumstances of scientists and does not need to claim transcendent truth. We can't help revealing our notions of biologi-cal design in our other creations, just as we can't help showing ourselves in our practice of biology. Probably the most common and intuitively ap-pealing way of conceptualizing natural design, in fact, is by analogy with our own activity. To acknowledge our part in constructions of Nature is to accept interaction as the generator of ourselves and of our interrela-tions, of knowledge, and of the world we know. Both we and the world are expressed in this dance, even as we are created and know ourselves in it.

If any of this makes sense, then the perspectives of the scientists who are exploring alternative approaches are extraordinarily precious. Their work shows us selves both responsive and response-able (and so, responsible) to a different conception of Nature, and thus to a different conception of ourselves. If there is special value in their work, and I think there is, it is perhaps not so much due to their having found the one true essence of Nature, but rather to their having developed and enacted a "knowledge of the larger interactive system"—what Gregory Bateson calls, simply, "wisdom" (1972:433). Such knowledge is not in any absolute sense truer than the partial knowledge of cause-effect arcs that Bateson says our conscious purpose cuts out from the loops and circuits of larger systems, but it may well be crucial to our particular, nonabsolute contingent, and increasingly endangered existence.

CONCERNS, CARES, AND GENERATIVE COMPLEXITY

What sorts of concerns are expressed in these alternative ways of questioning Nature? In what may be a foolish attempt to speak for those of my colleagues who are turning their science to sustainable agriculture, energy-efficient technology, the use of biological communities to degrade waste, and other important tasks, I will venture a few guesses. One thing I see is worry, both about the excessive power of traditional science and about limitations on its power. Science's ability to predict and control (the twin goals of contemporary science—whatever happened to the ability to understand?) often seems inadequate to the cascade of unintended consequences that frequently follows technological advance. These two, excessive power and inadequate power, are not contradictory. Both are aspects of our embeddedness in the world, an embeddedness denied by conventional accounts of objective scientific knowledge. As an example of this denial, consider the idea of objective truth as truth that is grounded in no ground, no body, and no place. Quantity of power, furthermore, is not necessarily the same as adequacy or appropriateness of power. A little delicacy and discernment can go a long way.

In addition to worries about power and vulnerability, I sense in these scientists a certain discontent with the feeling of estrangement that seems to have been fostered by establishment scientists' obsession with detachment, isolation, and independence. Metaphors of engagement, connection, and interdependence crowd their language, as they do mine, as we struggle for different ways of knowing and being in the world. This emphasis on interconnection is related, in fact, to my points about power and vulnerability. Attending only to the intended consequences of powerful techniques without attending to the consequences of those consequences is just what allows mainstream science to advertise power without giving comparable time to danger and mishap. Often associated with the paralyzed dismay that can accompany our glimpses of what Donella Meadows calls "Awful Interconnections," these metaphors of wholeness can also show us that "solutions are as interconnected as problems" (1988:16). Solutions, in turn, point to other problems and solutions, not only technical but psychological, cultural, and economic as well. There is thus a moral-aesthetic-political dimension to this preference for inclusion over exclusion, the personal over the impersonal, mutuality over domination, systems of influences over single causes, openness over closure, loops over lines.

This dimension exists whether we recognize it or not, and it is a source of difference in our approaches to science. Differences, however, do not necessarily preclude common goals. I suggest that we try to respond to Haraway's challenge to forge a coalition of diverse tongues and visions, generative of the kind of complex and multiple truths required by our complex and precarious designs with, of, on, and in Nature.

9 Bodies and Minds:

Dualism in Evolutionary Theory

The first wave of controversy over sociobiology has subsided. Cautious advocates make multiple disclaimers: Sociobiology is not racist/sexist/ reactionary or anticulture. It does not require genetic determinism or nature-nuture dichotomies. It does not show us what characteristics are inevitable or desirable, or even reveal the limits to human possibilities (Alexander 1987; Crawford, Smith, and Krebs 1987; Daly and Wilson 1988; Gruter and Masters 1986). Meanwhile arguments about evolution are heard with increasing frequency, not only in biology but in other disciplines and in the popular media. Volumes on sociobiological approaches to political science, psychology, and the law have appeared (Beckstrom 1985; Crawford et al. 1987; Gruter and Masters 1986; MacDonald 1988a). Talk of "inclusive fitness" is common, and the assumption that organisms, including humans, are constantly jockeying for reproductive advantage raises fewer and fewer eyebrows.

The Domestication of Sociobiology

Certainly the sociobiologists' assurances are meant to show that the fuss of the 1980s has been misguided. Bad old words such as *instinct* and *innate* have largely been replaced by substitute phrases that are supposed to be innocent of unwanted implications. Thus we hear of evolutionarily derived or canalized behavior, adaptive central tendencies, evolved motives, and facultative adaptations (ones that take different forms under different conditions, thus demonstrating that adaptiveness does not always entail developmental or behavioral rigidity). Sociobiology seems to have become thoroughly domesticated, and a casual observer might conclude that all is well.

New formulations, however, often carry the same sorts of problems that plagued the old, and ambiguities and questionable inferential links sometimes persist despite authors' best efforts to eliminate them. In fact, some scholars insist that there is still a need for the old concepts (Fuller 1987, on "something like" instinct being passed on in genes; Silverman 1987). Such expressions of intellectual nostalgia suggest that terms are tidied up more easily than thoughts. Although some conceptual change has occurred, thanks to both evolutionists and their critics (overlapping groups, incidentally), we still need to clarify just what an evolutionary account requires, and what it can and cannot tell us about ourselves.

The issues raised in this chapter are not confined to sociobiology. Some are not even specific to evolutionary inquiry, but involve pervasive ways of thinking about ourselves and others: about evolution, the body, and "human nature." I start with a brief discussion of some relations between the body-mind problem and the nature-nurture dichotomy. I return to two variants of genetic determinist thought, the beliefs (1) that evolution produces fixed universals and (2) that the genes determine a fixed set of alternatives. Then I examine a rather more subtle matter—the belief that evolution necessarily produces "selfish" creatures relentlessly pursuing reproductive advantage. While the former issues turn principally on the nature of development, this last one also involves the moral meaning of our acts.

One of the legacies of the nature-nurture dichotomy is that anyone criticizing one of the opposing positions will be seen as advocating the other. If one voices skepticism about some "biological" interpretation, then, one is assumed to be an environmental determinist, and vice versa. This assumption is a trap, and it is better to dismantle traps than to step into them (or, for that matter, to set them for others).

Innateness: The Body in the Mind

The phrase "the body in the mind" comes from the title of Mark Johnson's 1987 book. George Lakoff and Johnson (1980) have investigated the "objectivist" view of mind as disembodied and abstract, arguing for an alternative that does not rest on a separation of mind from body (see chapter 3). I am sympathetic with this depiction of embodied thought, and these critiques of the opposition between subjectivism and objectivism show many affinities with my own analyses. I borrowed the title for a different reason, however: to highlight the association of innateness with the body.

Johnson and Lakoff wish, among other things, to heal the mind-body rift that has been so important in the Western philosophical tradition; Johnson's "body in the mind" is thus intended to signal an integration. I use the phrase, however, to indicate a *failure* of integration, a continued reliance on a set of ancient disunities, in which nature is associated with the body, and nurture with the mind. Calling some aspect of the mind "biological" thus invests it with "bodiness." This special status may depend on any of the meanings of *innateness*. If a trait has counterparts in other species, for instance, or appears universal in humans, or follows family lines in some population (shows heritable variation), it may be dubbed "genetic" or "biological." Other definitions are based on apparent survival or reproductive utility, the presence of identifiable morphological or biochemical correlates, apparent spontaneity or irrationality, presence in infants, or absence of obvious learning. By attributing part of the mind to nature, then, people place *the body in the mind*. Because it is a notion of body that is rooted in a dualistic framework, however, the result is not integration but only a more complicated dualism.

Scientists have not been completely oblivious to the weaknesses of the nature-nurture opposition; recall the disclaimers mentioned above. But disclaimers and adjusted terminology are not enough. Concepts can survive considerable turnover in lexicon. E. O. Wilson (1975) defines sociobiology as the study of the biological bases of all social behavior. Innumerable nonsociobiologists (physiologists, geneticists, anatomists, etc.) also consider such "bases" central to their work, so the continuing muddle about what a biological basis *is* can hardly be termed minor. (Add those who devote time to showing that something or other is *not* biological, and one has included much of the scholarly community.)[1]

All behavior requires a body (and therefore involves genes and environment, physiological correlates, brain structures, etc.), so nothing is served by partitioning the mind into biological and nonbiological pieces. Yet it is common for arguments about innate behavior to be buttressed by facts from neurology or biochemistry, as though there were some other class of behavior *not* mediated by neurons and chemicals (de Catanzaro 1987; Hoffman 1981). "Biological bases" are usually euphemisms for "instinct" and "innateness," and refer to the same disorderly range of phenomena. If the "nature" side of these dichotomies is not clearly delineated, furthermore, the complementary "nurture" side must be equally ill defined.

Asking about the role of learning in the development of some bit of behavior is not the same as asking about its phylogenetic history. But using *innate* to mean both "unlearned" and "shared by evolutionary relatives" obscures this fact. Similarly, whether or not behavior can be affected by any particular experience is quite independent of its survival value. Appearance early in life is not the same as imperviousness to outside influences. And so on. Once these nature-nurture questions are disambiguated, it should be clear that they are *different* questions, with different evidential bases, as we shall see below. They are not alternative ways of glimpsing a single underlying nature; rather, they reveal the diverse, sometimes conflicting meanings of "nature."

It is sometimes said that the division between nature and nurture needs healing. Similar remarks are made about the body and the mind. In neither case, though, are there two parts that need rejoining, like a broken dish. Both oppositions mislead by implying that their terms are of the same type, and that these terms are in complementary relation, defining some larger whole, the way complementary angles make up a right angle—that is, that behavior is partly innate and partly acquired, and a person is composed of a body and a mind. Once the metaphor of a partitioned whole is accepted, all sorts of oddities follow. An anthropologist may argue that a behavior pattern is cultural, not biological. Or, trying to convey a sense of unreasoning compulsion, a drug user may insist that the craving is *physical,* not just mental. A legal scholar may suggest that stepparents' biologically "programmed" tendency to abuse their stepchildren "may operate as a hard-to-resist impulse," and thus make the act seem less reprehensible than abuse by natural parents (Beckstrom 1985:133).[2]

There is nothing disreputable about research on brains and chemicals. What is problematic is the implication that innateness, however it is defined, has some privileged connection with those brains and chemicals.

After all, we learn and think by virtue of our bodies (the claims of some artificial intelligence theorists notwithstanding). The problem of the disembodied mind is not solved by embodying just the "innate" portions, for this would imply that other portions must be accounted for in some other way. However we gerrymander the outlines of evolutionary "nature," there will always be some residual chunk of acquired "nurture" hovering mysteriously in thin air, without a "biological base" to support it. But if natures are the result of the continuous nurture of developmental construction, we ought to object when we are told that nature interacts with nurture (or biology with culture, etc.), just as we do when we read about material body-stuff interacting with immaterial mind-stuff.

What Must Be: Fixed Universals

In addition to its dualistic overtones, the sheer ambiguity of the nature-nurture complex works mischief. When nature is identified by some criterion, it is easy to conclude that it is intractable or fixed, that *it must be*. This is probably the most common inference about "biology," and it is often unjustified.[3]

In this style of reasoning, it is usually *universal nature* that is supposed to be fixed. Martin Daly and Margo Wilson (1988:8–9) correctly point out that it is not only biologically oriented theorists who make claims about universal human nature; even "the staunchest antinativists" generalize psychological principles to the entire species. Daly and Wilson follow Symons (1987) in saying that the real point of contention between evolutionists and their opponents is the *specificity* of mechanisms, not their fixity, and that the nature-nurture controversy is a red herring. This is an important point, but part of the concept of human nature is that it is universal and at least relatively fixed. Insofar as the nature-nurture opposition is fueled by concerns about fixity, it is not so easily set aside. If it were not for the widespread conviction that "nature" is fixed, in fact, nature-nurture questions would surely not have generated so much heat over the years.

Many scholars have become aware of the inappropriateness of automatically assuming fixity whenever biology, evolution, or genes are mentioned. As I noted earlier, it is now common for presenters of evolutionary research to insist that the behavior they describe is not inevitable or immutable.[4] It is important to keep challenging that reflex assumption. Full assimilation of the lesson is hampered, though, by the associations described above, and by the fact that milder versions of the fixity argument perpetuate its spirit while softening its outlines. If the innate is not immutable, some authors tell us, it is at least hard to avoid or change.

M. S. Smith (1987:232, 244), for instance, implies that "strong cultural pressures" are needed to overcome "evolved dispositions," because evolutionary theory tells us that "some aspects of human behavior are easier to change than others." Apparently relying on the opposition between biological emotion and (nonbiological?) reason, Gruter (1986:277) warns that laws prohibiting acts that *"fulfill survival functions"* are unlikely to be effective because strong emotion is impervious to reason. According to Alexander (1987:218), it will be hard to persuade men to use contraception because it reduces their reproductive opportunities. It will also be hard, he thinks (Alexander 1979:233), to achieve altruism within groups without intergroup hostility because competition among groups has been crucial in evolution.

As we have seen, there are other variants of the idea that the "biological" is hard to change, including the notion that biology sets limits on potential. The point is that they *are* variations on a theme, and are as untenable as claims of strict immutability. Notice that I am not making a case for "infinite potential," whatever that might mean. Some behavior may be impossible to change. But there is often no simple way of finding out how hard something will be to influence, short of trying. We are sometimes surprised by the outcome of our efforts. Responsiveness may well depend on the larger context; even patterns that were stable over evolutionary time may not be so in new environments, and vice versa. Many theorists, in fact, have emphasized the important role of behavioral innovation in bringing about evolutionary change (P. Bateson 1988a; Gray 1988).

A desire for simplicity and certainty is understandable, particularly with respect to socially important issues such as violence, competition, and hierarchy, or family and gender relations. It is risky, however, to conclude that some aspect of contemporary life is inevitable on the basis of a phylogenetic or a functional argument. It is seldom certain that a statistical phenomenon like the allegedly greater frequency of child abuse by stepparents than by natural parents (Beckstrom 1985; Daly and Wilson 1988) can be changed. Does the addition of an evolutionary explanation warrant more pessimism than the data on distribution alone (including cross-cultural and historical distribution)? That would be the case only under the assumption of a single fixed nature.

What Must Be: Fixed Variations

The 1980s and 1990s have witnessed increased attention to behavioral *variations* rather than universals (Caro and Bateson 1986). Facultative adaptations are morphological or behavioral variants that appear only under certain circumstances and seem advantageous in those circumstances. They are often presented as proof that evolutionists are not concerned exclusively with the immutable, and yet they are sometimes conceptualized in quite rigid ways.

Thornhill and Thornhill (1987:280) hypothesize a tendency of human males to rape when they are at a competitive disadvantage. The Thornhills suppose this to be a universal tendency controlled by a "single general genetic program" that responds to social conditions. They do not claim that every man always rapes, but rather that every man "tends" to do so under circumstances in which the reproductive deck is stacked against him. The act is not inevitable, in their account (it will occur only in certain situations), but the "programmed" behavior-situation connection seems to be. This is a slightly different sense of *what must be,* but the supposedly pan-human distribution of this "tendency" shows the conceptual kinship between this account and traditional universality arguments. Where earlier theorists assumed the appearance of some behavior on the basis of an evolutionary argument, this view assumes the situation-behavior *link.* Genetic programming, however, is not a developmental mechanism: It does not *explain* reliable development, it *renames* it (when it is invoked after the fact) or, more dangerously, *assumes* it (when it is used predictively). "Programming" serves, then, as either pseudoexplanation or prognostication.

That such prognostication can have serious consequences is seen in Shields and Shields's (1983) recommendations for reducing the incidence of rape. After presenting an evolutionary explanation of rape that differs from the Thornhills' in emphasizing male hostility, they state: "[We] assume the existence of a polygenic substrate that generates a *closed* behavioral program linking hostility and female vulnerability with forced copulation. We perceive this connection as "hard wired" and the genetic substrate encoding it as essentially *fixed* in the human population. From this perspective, we view rape as a human universal and on *this* level, probably recalcitrant to reprogramming via individual experience" (Shields and Shields 1983:123).

The model also includes an "open" program for "assessing female quality and vulnerability" (p. 123). The authors state that their model of an innate threshold for rape implies that changes in men's attitudes or women's status would do little to reduce rape. Although the authors disapprove of sexism, they say there is little chance of changing the rapists themselves: "If the evolutionary hypothesis of a genetically fixed substrate strongly predisposing men to rape under the appropriate proximate conditions were true, then rehabilitation via education might be difficult, if not impossible." Instead, they advocate sure and severe punishment, which would not only deter rape, but if "sufficiently severe . . . could act as artificial directional selection against lower thresholds" (Shields and Shields 1983:132–133).

There seems to be a confusion here between (a) fixation of a gene or genes in a population and (b) rigid developmental fixity (unresponsiveness to variation in developmental environment). These two are quite distinct, of course. What links them is not empirical or logical necessity, but rather the very system of beliefs about intractable genetic nature that we are examining here. Concerns about fixity are not confined to evolutionary thinking. The point is that the assumption of static nature is very pervasive, and is not fully eliminated by invoking the environmental sensitivity of facultative adaptations. The nature-nurture dualism threatens to prejudge empirical questions that should be investigated on their own.

What Really Is, or, What Do People Really Want?

Biology is often thought to define the timeless truth that underlies shifting appearances. Consider the spatial metaphors in the preceding sentence: Biological reality lies below psychological appearances; it is the foundation for everything else. In this vision of a layered reality, the biological is given special status. Related to this is the postulation of real, evolutionary goals underlying everyday motives. The conceptual play here is between evolutionary *consequences* and human *intentions;* by sleight-of-thought, a history of natural selection is transformed into a contemporary psychological truth. Although we frequently invoke the distinction between "ultimate causes" (conditions prevailing over evolutionary time) and immediate "proximate causes," we disagree on just how to keep them apart. One problem is that the distinction tends to collapse: Ultimate causes are transmuted into proximate ones, often in the guise of genetic programs. Thus, in some human sociobiology, the evolutionary past is present in the innate selfishness that is supposed to underlie all behavior.

It is often said that knowledge of mechanisms (i.e., proximate causes, including psychological processes) is unnecessary to evolutionary arguments — that organisms need not "consciously" try to maximize fitness, but only *act* as though they were doing so. This misleads even as it instructs. Organisms may do things that have, to some degree, been associated with fitness over evolutionary time, but fitness maximization itself need not be the proximate goal. The organism may well be tracking (adjusting its behavior to) some other variable. To say that organisms are not maximizing fitness *consciously,* furthermore, is to invoke an *unconscious,* which is conventionally seen as the proximate cause that *really* explains behavior. For instance, Hrdy and Hausfater (1984) say that human parents calculate, consciously or not, the costs and benefits of infanticide. (See also Shields and Shields 1983; and Thornhill and Thornhill 1987, on males' calculation of costs and benefits of rape.) This idea of unconscious reckoning fits very well with the idea of evolutionary causes "underlying" more superficial phenomena, perhaps in the form of genes manipulating organisms (Dawkins 1976), as well as with the Freudian viewpoint about hidden psychic forces. And so we find declarations about what people are *really* trying to do, regardless of what they think or say; for example, M. S. Smith (1987:235) tells us that, whatever they may think, parents' "real goal" is fitness maximization.

Pointing to the unconscious implies that humans are self-deluding rather than hypocritical. Alexander's characterization of the conscience, however, at least suggests the latter. It is the "still small voice that tells us how far we can go without incurring intolerable risks"; it "tells us not to avoid cheating, but how we can cheat socially without getting caught" (1979:133). Even the usually cautious Daly and Wilson, who warn that "evolutionary psychology is not a theory of motivation" (1988:7), speak of the importance of feigning sincerity to manipulate others, and dismiss "mystico-religious bafflegab about atonement and penance and divine justice" as "actually a mundane, pragmatic matter" of curbing others' competitive strivings (pp. 255–256).

Whether these writers are denying the moral import of our choices on the basis of our evolutionary history or telling what (amoral) choices we are *really* making (see also Kitcher 1985), their persistent characterization of humans as devious egoists goes beyond what is required by evolutionary analysis. They caution us not to treat proximate and ultimate causes as alternatives (Alexander 1987; Daly and Wilson 1988), but if they are telling us that we cannot be both moral beings and evolved creatures, they come perilously close to doing just that.

As we have seen, for a behavior pattern to confer a fitness benefit, it need not have just that benefit as its goal (recall Daly and Wilson's 1988 caveat about motivation). It is enough for its goal to be associated with survival and reproduction. By "goal" I mean an empirical correlation—What does the behavior track? To what does the organism adjust its behavior? Even the much-cited hypothesis of human incest avoidance surely involves no concern on the part of the actors about inbreeding, only some disinclination to mate with those with whom they have been raised (see references in van den Berghe 1987). Neither an overt nor a hidden (unconscious) concern about incest need be involved; it would be peculiar to claim that a young woman falling in love with someone she has just met is "really" trying to reduce the dangers of inbreeding and so enhance her fitness, as though careful psychoanalysis would reveal these hidden desires. Such a claim confuses the idea of natural selection with the physiological and psychological processes of individual organisms. Because the organisms we see are those whose ancestors developed and behaved in ways consistent with leaving viable progeny, the organisms themselves (including ourselves) are now said to be "trying" to leave offspring.

Evolutionary Tough-mindedness

Cynicism can be attractive. It seems sophisticated and intellectually advanced, signaling a hard head, not a soft heart. Some evolutionary writings can be seen in this light. They also exemplify a tendency within evolutionary theory to reject anything but egoistic individualism as a romantic leftover of sentimental group selectionism.[5] Neo-Darwinists frequently present themselves as tough-minded scientists capable (unlike social scientists, not to mention people in the humanities) of facing unsavory biological facts. This self-conscious tough-mindedness, rather than any particular politics, may help us understand some evolutionists' depictions of human affairs.

Thus MacDonald (1988b:140) characterizes sociobiology's lesson for psychologists as the expectation that humans will maximize their self-interest, deceive themselves and others, and believe what it is in their interests to believe. Moral reasoning, he says, is epiphenomenal, merely "masking self-interest," and adds that this insight was already attained by the behaviorists. This last point is important; egoistic individualism did not originate with sociobiology. It is a dominant theme in Western thought (Caporael et al. 1989; H. Margolis 1987; Schwartz 1986). Before sociobiology appeared, the rational calculator of preferences could be found in economics. The goods sought by economic man had no *inherent* value; value was conferred by each person (Schwartz 1986:57–60). Sociobiologists have now supplied the inherent value: reproductive advantage. More mundane desires for wealth and power are sometimes said to motivate us, on the assumption that they in turn maximize fitness. This lack of clarity about goals and intentions, conscious and unconscious, reveals confusion about just what claims are being made about psychological processes. Whatever uncertainties exist, however, these theorists seem to have few doubts about the fundamental selfishness of our species.

Alexander (1987:223) states that the social sciences are frequently infected with moral concerns, but that such intrusions are foreign to biology. I disagree. Nor, as I maintained in chapter 8, do I think we should strive for a completely insulated science, even if that were possible (Longino 1990). In *Habits of the Heart,* a sensitive examination of American conflicts over individualism and social integration, Bellah et al. (1985) provide some background for considering the ways we attempt to thrash out these issues, not just in our private lives, but in our science as well.

To the everyday claim that humans are basically selfish, some current evolutionary arguments add a ratification by natural selection, along with the redefinition of altruism in terms of genetic "selfishness"—terms that make the "self" in question not a person or any other organism, but a snippet of DNA, whose goals we supposedly pursue. We do not need evolutionary theory to tell us that humans can be ruthless, manipulative, and greedy. We should not let it tell us that humans are necessarily dishonest or deceived when they behave well (Kitcher 1985; Oyama 1989). If evolutionists wish to convert the thoughtful and cautious, not just the most eager, they must put their own theories in order—especially considering the harsh words some of them have for the social sciences. If social scientists wish to benefit from the new biology, they must make sure it is not a biology that commits them to the very dualist framework that has kept biologists and social scientists apart for so long. (For a careful attempt at integration without oversimplification, see Hinde 1987.)

Passion and Calculation

People have tended to associate biology with the body, the passions, the irrational. This venerable network of meanings has helped generate and maintain notions of innateness as an animal (often beastly) nature within us humans (Klama 1988), at war with reason and capable of overwhelming it. This opposition between reason and instinctive passion is evident in some of the works cited above. With the growth of sociobiological theory, however, a fascinating, if subtle, change has occurred: Long considered our best defense against the beast within us, rational deliberation is being appropriated to do that animal nature's work. As we have seen, certain theorists suppose us to be forever doing the bidding of our selfish, fitness-hungry genes. Their machinations conveniently unconscious, these calculating prodigies are said to produce, and explain, our behavior. Regardless of what we *think* we are doing, we are *really* serving our genes' interests. It seems that the biological beast has commandeered the very rationality that used to keep it in check.

Humans have long considered reason to be what most clearly separates us from other creatures. In our more expansive moments we have read both ontogeny and phylogeny as chronicles of the triumph of the intellect over emotion, the mind over the body. The id-ridden infant comes to be ruled by the reasoning ego, and small-skulled ancestors gave way to brainy humans. Science fiction imagines still more highly evolved creatures than ourselves, with huge, domed crania, or even beings of pure intellect who have left the body behind. Our notions of science, too, emphasize the liberation of reason from desire. As sociobiology has been domesticated by becoming part of common knowledge, however, reason has been domesticated in a different sense: Hitched by inclusive fitness theory to the plow of fitness, it has been given a new destiny—to cultivate the fields of genetic interest forever.

To those who sense the dangers of attributing purpose or volition to the genes but nevertheless champion genetic programs, I would suggest that those dangers cannot be avoided by piecemeal adjustments of vocabulary and argument. Fuller (1987:149), for example, says the program metaphor is not useful for explaining behavior, but then goes on to describe animals as "programmed by their genes to assign a higher value to the potential immortality of their genes, than to their own immediate welfare" (p. 169). A computer program is an expression of its creator's will, and as we shall see in the next chapter, to speak of genetic programs as the genes' way of controlling a body is to invite confusion over whose will actually counts in assessing human action. Faced with attempts to replace moral discourse with the language of genetic calculation, we ought to think very carefully about just what an evolutionary analysis does or does not dictate about psychological processes, and about just whose reasons we are willing to take seriously.

10 How Shall I Name Thee?

The Construction of Natural Selves

There is a growing literature on the social character of knowledge, on meaning as negotiated in social interchange, and on the importance of interpretation in knowing (see, e.g., K. J. Gergen 1985; Giddens 1987; Harré 1991; Henriques et al. 1984). In questioning the possibility of pure knowledge of the world, these authors challenge the vision of isolated subjects that informs much of scientific and everyday life. Several themes recur in this heterogeneous, contentious literature. The mutual reinforcement of cognitivism, the Cartesian model of knowledge, and mainstream science, for example, are frequently mentioned, as is the need to reconceptualize human subjectivity and recontextualize science, to make explicit the subtle and not so subtle ways in which science reflects and influences less formal ways of thinking about ourselves.

Descartes' Ghosts

COGNITIVE AND DEVELOPMENTAL CONSTRUCTION

In the traditional academic hierarchy, psychology often seems to occupy an intermediate position, somewhere between biology and the humanities. Biology speaks of behavior in terms of the body, largely in the language of causes, while nonscientists tend to use the language of mind and reasons. It is psychologists' uneasy task to mediate between these realms, between the biological discourses of evolution, function, mechanism, and physiology, and the political and ethical discourses of persons and acts (N. Rose 1985:13; Venn 1984:127). Psychology serves, in short, as a sort of disciplinary pineal gland.

Building on a distinction between necessary nature and contingent nurture, however, psychologists frequently oscillate between the two realms, patching together an unintegrated combination of the biological and the cultural, the physical and the mental, and, as we shall see, even the determined and the free. Insofar as it is committed to being scientific, which today often means being biological (and, increasingly, cognitivist, in the sense of thinking in terms of information-processing mechanisms modeled on computer technology), psychology affords less and less room for human subjects. Indeed, part of the mission of modern science is the replacement of mentalistic explanation with mechanistic accounts, of intentions with causes. This has left the science of mind in a confusing position. Biological explanation sometimes seems to threaten the realms of personal experience and responsible action that are implicitly associated with the psychological level, and various attempts may be made to salvage the subject, often by invoking social factors. Sometimes, though, psychology seems to do away with persons altogether, as when it conceptualizes humans simply as the products of genetic and environmental (including social) causes.

Developmental psychologists, whose job it is to chronicle the ontogenetic history of this divided being, have built their observations and theories around the innate-acquired distinction (Johnston 1987; Riley 1978). Because giving an account of development is a powerful way of characterizing the present, furthermore, developmental dualism tends to inform our attitudes toward the persons who result. Whenever a person's true nature is distinguished from accidents of his or her life history, or when potential for change is assessed on the basis of "biological components," or, as in chapter 9, when real, evolutionary goals are declared to lie behind apparent ones, nature-nurture distinctions are deployed to delineate a fixed reality behind the visible world. Accordingly, a boy described as genetically predisposed to criminality may be treated quite differently from one who is thought merely (!) to have fallen in with the wrong crowd: The developmental story is also a pronouncement on what is fundamental in the present—some inner "propensity" that must be suppressed or channeled. Similarly, people who believe aggressiveness to be humanity's inborn evolutionary legacy may have different expectations about the possibility of reducing violence from those with other beliefs. Or people's view of altruism may be colored by the conviction that genetic selfishness provides the real explanation of any behavior (Oyama 1989).

Scientific naming can influence everyday understandings after all, though generally not in a simple or direct way, and these in turn can enter into the developmental construction of persons. These two senses of *construction,* cognitive and developmental, appear repeatedly in much of the discussion that follows. Based as it is on the developmental systems perspective, my account stands in contrast with many uses of *construction,* including many social constructionist writings. Mentioned in the Introduction, this contrast is further explored below.

Cognitive and developmental construction can be related in several ways:

1. Cognition is a way of interacting with the environment, and knowing changes the knower, for the short or long term, so cognitive change can be seen as developmental in itself.
2. People respond to the reality they construe, so their understanding of their surroundings influences the salience and significance, and thus the developmental impact, of those surroundings.
3. People's cognitive constructions of themselves affect, and are affected by, other aspects of their development. A political, psychotherapeutic, or religious conversion, for instance, requires a *reunderstanding* of oneself and one's relationship to the world, and thus, changed ways of relating to it. One becomes a different person in a different world.
4. Others' constructions of us (including psychologists') may also influence what we become—they are parts of our social environments.[1]

These links between cognitive and developmental construction are themselves linked. Such mobile, complex, variable processes are not captured by schemes that cast either insides or outsides in a primary role; "expression of genetic potential" is as inadequate as "internalization" at describing the formation-in-interaction of developmental emergence. My preference is to construe development or ontogeny—synonymous in my usage—very broadly, by referring to the latter word's roots: "the genesis of being." This coming into being continues throughout life, and is as various as individuals are.[2]

THE BIOLOGICAL AND THE SOCIAL,
THE CONTROLLED AND THE CHOSEN

Biological nature tends to be thought of as fixed, and so beyond individual or social control. An example can be seen in Berlin's (1988) discussion of sexual aggression, in which the author wonders whether moral talk is appropriate for people who struggle against their "nature," their "biologically based drives." He suggests that some sex offenders may be incapable of responsible action rather than evil. Berlin cites cases in which such offenders were found not to be criminally responsible: To be a full person before the law, it seems that one must not be too biology-ridden. Here we see one way in which the languages of science and the larger society, often thought to be comfortably separate, rub uneasily against each other. The law requires autonomous beings; biological explanations seem to compromise autonomy. But perhaps, rather than asking whether behavior is caused or chosen, we should seek an understanding of moral action that includes the causal relations of which science speaks. Later we shall consider the possibility that moral accountability is more a matter of one's position in the social world than of transcending or denying causal constraint, biological or otherwise.

Biologists did not create the attributional complex that allows us to think so easily of inner forces compelling unwilled activity. A commonsense background of beliefs about responsibility, self-control, and other-control informs scientific thinking more than most scientists would care to admit—the moral power of the "whisperings within" (Barash 1979) style of sociobiological talk derives from a rich network of ways of construing our actions. Whenever biology, however defined, enters the discussion, the move from person-as-subject to person-as-object is facilitated. Consider the tendency to adopt an exculpatory attitude toward disturbing behavior when it is attributed to disease. As with the sex offenders, responsibility may be seen to be incompatible with biological explanation.

We constantly acknowledge and deny authorship of our behavior. We also attribute responsibility to others (K. J. Gergen 1990); this is part of the way persons can be constructed, maintained, threatened, and changed. But the opposition between the biological and the environmental, frequently invoked in these exchanges, can enter into people's very understandings of themselves (cognitive construction), and so into their constitution as persons (developmental construction).

Biological causation is not the only factor that diminishes personal responsibility. The language of environmental causes apparently eliminates agency altogether. Although we may be more familiar with the scenario in which a relieved parent hears that a child's heretofore inexplicable behavior is due to its inborn temperament (which lets both parent and child off the responsibility hook), we should also recall B. F. Skinner's (1971) declaration that the language of personal responsibility becomes misleading once the response contingencies governing behavior become clear. As biological determinism's mirror image, environmental determinism shares too much with its opposite to be of significant use in thinking about development (or morality; Oyama, 1990b). Despite the fact that environmental determinism has often been used to combat biological determinism, furthermore, it has had the paradoxical effect of further diminishing personal agency by adding more causes to drive out reasons. Insofar as even recent, more sophisticated kinds of social constructionism rely on what I have termed "developmental dualism," they tend to invite a complementary incursion on personhood: control by society, not biology. As human agents are rendered impotent, however, other agents are invoked to make and control them. In fact, people speak of being programmed by their parents or culture as well as by their genes, with similar suggestions of reduced accountability for what one might call "the not-me in me."

The persons-situations debate in social psychology shares this inside-outside opposition, as well as the difficulties of integration. Edward Sampson says, for instance, that "situations structure, organize, and regulate knowledge to more of an extent than individuals structure, organize, and regulate situations" (1991:52; see also Shotter 1984:123). But making the story more social does not make it less dualistic. The notion of a preexisting environment that shapes persons, and to which they must adapt, reinforces the inside-outside dichotomy. Invoking deterministic situations, then, no more resolves this antinomy than environmental determinism resolves developmental dualism.

THEORY AS PRODUCTION

In *Social Accountability and Selfhood,* John Shotter (1984) criticizes psychology's notion of persons. His primary concern is the Cartesian worldview, in which people are set apart from the world and mind is set apart from matter.[3] In presenting his notions of socially embedded personhood, and of responsible action as accountable, committed, and always occurring in a moral order, Shotter elaborates a vision of life as being-in-the-world, and of the world as "a growing world of form-producing processes" (p. 54). He denies that science is, or should be, insulated from the value-drenched world outside, and shares others' concerns about the individualism fostered by current science: Our ways of experiencing are influenced by our "ways of talking" (pp. 173–174), including scientific talking. Those who have raised these and related issues desire a remoralization of life—a recognition that morals are implicated in the very fabric of our social (that is to say, our total) being.

Notice that as conceptions of cognition, sociality, and development are broadened and related, they become part of the same process (Valsiner 1991). I would call them all biological, not in an exclusive sense that distinguishes them from other, nonbiological aspects of persons, but in the inclusive one of "pertaining to life." Human biology is then not a matter of individuals with fixed internal natures, but of changing natures that are a function of reciprocal relations with environments that always have a social aspect.

Developmental psychology would seem an especially fertile ground for investigating such issues. Like evolution, development can be read as an origin myth, a narrative that characterizes the present as it names the past. Tales of our movement from a past "there" to a present "here" tell us what we are by showing how we came to be (Oyama 2000, Riley 1978). Developmental psychologist Jerome Kagan (1984:28) observes that American psychology's antihereditarian emphasis between the two world wars was related to hopes for social equality. Active babies gave way to passive ones as external influences became primary. If contemporary concerns about morality persist, Kagan suggests, developmentalists may return to late-nineteenth-century preoccupations with conscience, choice, and volition.

This is no blind prognostication, but rather a comment on the burgeoning critical literature cited at the beginning of this chapter. Although Kagan's explicit attention to the influence of the larger societal context on scientific thought is welcome, he devotes little space to the impact of scientific constructs on our everyday experience of ourselves and others. Yet, as Rappoport (1986:175) says, "when we re-name important aspects of our world, we transform ourselves." There is a sense in which we simultaneously name ourselves when we name the world—but these namings are not achieved in isolation. They are open to challenge and revision. Many battles have been fought over what something should be called, and one reason is that naming—talking—is not only an action itself, it is related, albeit in complicated ways, to other actions.

Wielding social constructionist accounts of the mind to counter biologists' or cognitive scientists' individualistic ones is not the most fruitful way of opening up space in which to think about experience and choice. It may be that some of these ways of theorizing human action, often attempts to rescue it from biology by invoking learning, stem from unhelpful ways of thinking about developmental causation. Several questions about the conceptualization of human experience and action arise from these themes:

1. Why should there be such striking parallels between depictions of genetic and cognitive processes?
2. Are these parallels related to more general difficulties in thinking about causes and reasons and the origins of organized form?
3. Is there a way of conceptualizing emergence that does not involve the same pitfalls?

Who's in Charge Here, Anyway?

HEADS AND CELLS

Scientists sometimes discover the Cartesian subject not only in the human head, but also in the cells that make up that head (as well as the rest of the body). Why should there be such striking parallels between the problem of the subject in developmental psychology and the problem of conceptualizing biological, especially genetic, processes? In both cases, a single source of organizing (information-processing) power seems necessary, so that the cognitivist subject and the homunculoid gene play similar roles in the dramas of cognitive and ontogenetic formation.[4] The seventeenth-century scheme of pure individual knowledge so important to many present-day views of mentality and the split between immaterial mind and material body inhabit contemporary constructions of ourselves. When psychologists drive the homunculus-in-the-head from one part of their theories, it tends to crop up in others (Henriques et al. 1984:17; Shotter 1984:126). Similarly, developmental biologists take pride in having eliminated the curled-up preformed manikin from the germinal cell, yet have no problem believing in microscopic intelligences in the nucleus that plan and control the creation of the organism. The much-criticized presocial, rational subject not only resembles the macromolecular subject, it is thought to be the product of the activity of that subject. Thus we are haunted by Descartes' ghost, not only when we directly theorize humans, but also, as we have seen, when we describe their development. If matter is inert, then organized processes can be explained only by reference to a structuring intelligence. But if it is interactive (think of any chemical reaction) under changing, interdependent constraints, such outside direction is not needed.

COGNITIVE AND GENETIC SUBJECTS

These problems may be implicated in psychology's difficulties in conceiving human subjectivity. Mental processes seem to require Cartesian ghosts to compare current input with memory and perform other operations in that hospitable space behind the eyes. The language of information processing is supposed to explain how the outside is brought inside (Henriques et al. 1984:19–21). John Searle's (1980) infamous man in the Chinese room, simulating knowledge of Chinese by shuffling correspondence rules written in his own language, is an argument against attributing minds to computers. (Searle's point is that the man, lacking semantics, does not understand Chinese, even though he can respond "appropriately" to input in that language, and that computers do not understand things either.) This man-in-the-room, deliciously evocative of the homunculus-in-the-head, is forever out of touch with the world outside even as he frantically processes input and cobbles together output so effectively that he seems, from the outside, to understand Chinese.

If it has been difficult to speak of mental life without invoking a man in the head, the situation with the genetic "subject" is not so different. The complexity and regularity of species-typical developmental courses are attributed to something internal to the organism, something that contains plans for its future form and function and that controls the realization of those plans. The only satisfactory candidate for cellular chief executive officer seems to be the gene. Genes can have their effects only by entering into biochemical processes, which are largely a matter of the spatial relations and reactivity of molecules, and which depend on precise and complex conditions. Yet, certain features make them well suited to be ghosts in the cellular machine: their encapsulation in the nucleus, their apparent ability to effect change without being changed. There is even a secret code. (Compare the genetic code for matching bases and amino acids with the rulebook Searle's captive uses to manipulate Chinese symbols correctly.)

The downplaying of ecological embeddedness and context dependency in biological accounts of development, easiest when outcomes are reliable and uniform, is similar to the Cartesian emphasis in psychology. Indeed, Kagan (1984:xv) explicitly likens cognitive "managers" regulating intellectual functioning to "the watchful eye of the nuclear DNA." I coined the term *homunculoid gene* precisely to convey the way we analogize certain natural processes to our own actions (depersonalizing ourselves as we do so). Thus biologists, laboring to produce demystified causal explanation to replace intentional accounts, have created something of an anomaly. The more astounding the phenomenon, the greater the temptation to supplement their descriptions with some overarching intellect. They reanimate their mechanical descriptions with an intelligent "soul," in short, a ghost. Once one accepts a breach between passive material body and active nonmaterial mind, it is difficult to have one without the other, especially when faced by the intricately coordinated doings of organisms.[5] It is thus far from sufficient to confront these phantoms in developmental and social psychology, where the absence of a satisfactorily fluid and rich conception of subjectivity has been decried for some time. Rather, we need to mount a broader effort, one that recognizes the relationships between our notions of subjectivity and our descriptions of process in the world around us. If Descartes' ghosts are mischievous ones, they need to be exorcised—not only from our heads, but from their other places of refuge as well.

Causes and Reasons

My second question was whether these parallels between depictions of genes and minds can be related to more general difficulties in thinking about organized processes and forms in the world, and perhaps even about the old puzzle of how to fit intentional action into a deterministic world. As we have seen, developmental theorizing is not the only thing that suffers when we cleave the world into the naturally given and the socially constructed. Insofar as the nature-nurture dichotomy maps, albeit imperfectly, onto the ones between causes and reasons, determinism and free will, involuntary and voluntary behavior, people are believed to be helpless to affect, and thus not accountable for, "biological" phenomena. This results in an odd situation in which persons and their genes become alternative agencies, responsible for different behavior. Or else causes and reasons are relegated to different realms (science and society, perhaps), but because of the many interactions between those realms, the separation cannot be maintained. This is related to instabilities in the psychology-biology relationship. Psychology's claim to be scientific is based partly on its ability to invoke the language, methods, and findings of biology, the neighbor discipline that supplies its "base" and that confers scientific respectability, an extra measure of reality and foundational importance to much that it touches. But with every flexing of reductionist muscle, biology also threatens to appropriate the very subject matter of psychology.

DISCIPLINES AND CAUSES

Symmetrical tensions between psychology and sociology (and some anthropology) reflect the complementarity of notions of genetic and environmental control, as we saw earlier. Giddens (1987:25) points out that sociology has often striven to show that human activity is governed not by individual will, but by social forces; and Shweder (1990:21) describes the disappearance of the person from ethnography as anthropologists countered psychological determinism with a determinism of the sociocultural environment. The reconceptualization of human subjectivity and agency would seem to require, then, a reworking not just of the psychology-biology boundary, but much else besides, including the boundary between the psychological and sociocultural realms. When this is accomplished, we may no longer think in terms of adjoining territories at all.

Shotter (1984:146) has said that although we attribute a projectile's path to general causal laws rather than to the moving object itself, we attribute people's motions to themselves. But, as I have noted, scientific explanation (aspiring to "general causal laws") tends to drive out intentional or agentive explanation, so that people are said to have been *caused* to do this or that, rather than to have acted for *reasons.* People also project intentions into other entities, endowing genes with plans or personifying social influences ("society needs compliant women, so it inculcates passivity in them"). Scientists caught in the act of anthropomorphizing in this way typically claim that it is merely convenient and has no pernicious consequences (Dawkins 1976, on selfish genes). Or they may deny that there is any metaphorizing going on at all (Dawkins 1986:111, on DNA as programs). But I submit that at least some of the perplexities experienced by psychologists—about development, the relations between causation and action, or between biology and psychology—are related to the fact that we seem to have difficulty thinking of certain complex processes without invoking some sort of prior intention to initiate and control them. A program, as noted earlier, represents the programmer's intentions.

When genes and environment compete for center stage, the person tends to be pushed into the wings (Shaver 1985:175). Insofar as biology stands in for fate and necessity, it can also be used to delimit responsibility. Although both genetic and environmental causes can drive out human choice, then, the former do so more consistently than the latter: Causes usually trump reasons, and genetic causes usually trump nongenetic ones. This explains the political potency of much biological argument. Consider the debates about differences between the sexes or races, in which fundamental nature is defined by pointing to biological indexes. Due to the incoherence of current notions of biology, however, the arguments cannot be definitive—not only are "all the data" never in on any particular question, but the *human significance* of the data is not at all clear.

Just as the child's contribution tends to get lost even in many "interactionist" (traditional, developmentally dualistic interactionisms, not the constructivist, systemic interactionism espoused here) descriptions of development, so does the organism tend to disappear from tales of gene-environment interaction. Recall also Shweder's point about disciplinary competition eliminating the person from ethnography. It is perhaps not so startling that what begins as a progressive impulse to rescue people from biological determinism often ends up subjecting them to an alternative determinism, leaving them as passive as ever (Henriques et al. 1984; N. Rose 1985).

BIOLOGY AND BLAME, INSIDES AND OUTSIDES

It is precisely because notions of causation and responsibility are so intimately entwined that these theoretical developments are also moves in ongoing arguments over whose fault it is that children are the way they are, or that the world is as it is. (Kessen 1979 mentions developmental psychology's propensity to blame mothers for errors in child rearing. We might add a more recent propensity to blame dead philosophers for our present difficulties.)

An article in a newsmagazine published in 1992 discusses the increasingly common sentiment that discrimination would diminish if the public would only realize that homosexuality is biological (Gelman et al. 1992). Genetic homosexuality is likened by one interviewee to left-handedness. Others cite the civil-rights protections accorded "natural" minorities; under the law, the authors say, "natural" means "immutable." (This is significant, given recent attempts in the United States to remove these protections by classifying homosexuality as an immoral choice.) Repeatedly stressed, in fact, are the supposed fixity of biological characters (so that parents need not fear homosexuals' influence on their children, whose sexuality is presumably determined before birth) and the concomitant relief from guilt for homosexuals and their parents alike (it's nobody's *fault*).

This welcoming of biological explanation is notable for reversing the usual strategy of denigrated groups, which, as I pointed out in chapter 7, has been to invoke nurture, not nature. In the face of a demand that one freely renounce sin, blaming biology has its attractions. (It may also comport with a sense of having discovered, not chosen, one's sexual orientation, or of having tried to change but failed. Given the prevailing framework, these situations are taken as evidence for "biology." See Kitzinger 1987 for alternative constructions.) The assumptions seem to be that parents (or homosexual models) are to blame for homosexuality if it is learned, the person him/herself is at fault if it is a vice, and nobody can be blamed if it is biological. But there is more than one way this move can play itself out.

Even if a biological argument is accepted, who can say that it will have the desired consequences? Renewed stress on "biological bases" could well propel homosexuality back into the realm of pathology, which it has never quite escaped. Having some aspect of one's personality labeled biological, we must remember, does not automatically confer legitimacy. Indeed, if legitimacy is lacking, biology easily sanctions therapeutic intervention, even against a person's will. Recall that Berlin (1988) thought certain sex offenders were not evil, but he had no objections to containing and managing them, which is just what we do with people who are not responsible for themselves. Disclaiming responsibility for oneself can be different from disclaiming responsibility for one's sexual preferences, but not everyone will see the difference.

As powerful as self-naming can be, transformation from an illegitimate/defective being to a legitimate/whole one is not the sort of thing that one can execute unilaterally. Like all such conferrals of identity, it is a complex process enmeshed in larger systems of power and meaning. To be sure, the biological turn may be seen as a declaration: "Here I stand; I can do no other." This is certainly a kind of moral stance, but I have doubts about its wisdom in the present instance. Martin Luther's statement was one of utter *voluntary* commitment. The biology argument, on the other hand, asserts that one is powerless. This seems a dangerous base for a claim to full personhood, and that is what I believe is at stake here. Additionally, replacing the moral model with biological determinism tacitly affirms that there are only two possibilities (crudely put, pure will and no will). This would seem to cut off the sort of inquiry that is called for here. Next, the strategy leaves unquestioned the assumption that sexuality is a fixed quality inside people, and that there are only two such qualities: heterosexual and homosexual. A move in a particular game of defining beginnings in order to construct an attitude toward the present, it does not question the coherence of the game itself. Finally, it would seem to make one's claim to be a person subject to empirical verification—and, of course, to potential disconfirmation—so that one's very selfhood would seem perpetually poised on the moving edge of "the body of data." John Money declares (in Gelman et al. 1992:48), "Of course it [homosexuality] is in the brain. . . . The real question is, when did it get there?" Was it there from the beginning or put there by some contingency of experience? Shotter, on the other hand, advises, "Ask not what goes on 'inside' people, but what people go on inside of" (1984:106).

One need not choose between speaking of insides and outsides, something that Shotter (1984:123) implies later in the same passage (although he sometimes uses the "outsides are more important than insides" style of argument as well). If insides are interdefined with outsides, we must ask what kinds of exchanges result in personhood. The invocation of essential nature, then, is a complex matter whose psychological, political, and cultural meanings must be worked out in the uncertain arenas of everyday life.

The legitimacy of one's self would seem to be more the point than the precise etiology given for it: Notions of biological feminine nature exist quite comfortably with discrimination against women. In fact, etiology becomes an issue when legitimacy is in question. (There seems no need to hinge *hetero*sexual rights on a developmental narrative.) Precisely because such outcomes are unforeseeable, multiple, and conflicting, I would not presume to advise against the moves in question. Indeed, if a broader swing back to biological explanations is in progress, the disempowered might be unable to halt it if they tried. More discussion might even encourage scrutiny of the whole issue, which could be a good thing. My comments, then, are meant to express my anxiety about these matters and my preference for questioning prevailing beliefs about biology, nature, and development rather than trying to turn a dangerously loaded dichotomy to homosexuals' advantage.

Forced Choices: Natural or Constructed Subjects?

Nature-nurture oppositions treat the genes and the environment as alternative causes of development, so questions like Is aggression biological or learned? or Are sex differences natural or cultural? tend to crop up. It is in this context that social constructionism has combated biological theories by emphasizing the nonnatural character of the self. These critiques, as important as they are, are based on the biology-culture dichotomy that I find so problematic. Having begun with the productive (and responsive) aspects of theory, we have inquired into parallels between genetic and human agents, and then into notions of causation and agency. Now I come to my last question: Is there an alternative to the dualistic schemes we have been reviewing? Or, as it was posed earlier, Is there a way of conceptualizing interactive emergence that does not involve the pitfalls we have been reviewing? My earlier sketch of development as constructive interaction, and of natures as emergent, mutable, and various, suggests that there is, and, not surprisingly, it points to the framework that is elaborated in this book. In this framework, subjectivity can be seen as both natural and constructed, at once wholly biological and wholly cultural.

DEVELOPMENTAL SYSTEMS

I take exception when Henriques et al. (1984:92) and Sampson (1991:16; see also Harré 1991; Shotter 1984:71) say that the subject is constructed, not natural. While acknowledging their points about sociality and historicity, I suggest that this formulation risks perpetuating some of the very inside-outside dichotomies they deplore. The same is true when Sampson (1991:211) says "there is nothing natural or inevitable" about selves (notice the connection of naturalness to inevitability). "Natural" biological persons are constructed, not only in the sense that they are actively construed by themselves and others, but also in the sense that they are, at every moment, products of, and participants in, their own and others' ongoing developmental processes. They are not self-determining in any simple sense, but they affect and "select" influences on themselves by attending to and interpreting stimuli, by seeking environments and companions, by being differentially susceptible to various factors, by evoking reactions from others. As Kagan (1984:279) says, "the person's interpretation of experience is simultaneously the most significant product of an encounter and the spur to the next."

To accentuate the irrelevance of the nature-nurture opposition to these processes, I have suggested a recasting of the terms. *Nature* then refers not to some static reality standing behind the changing characteristics of the phenotype, but to the changing organism itself. It is plural in a number of senses: Many "natures" (organisms-in-transition) constitute a species, rather than some single species essence, and an organism has as many "natures" as it has situational and developmental moments. *Nurture* becomes a cover term for all interactions that produce, maintain, and change natures. At the scale that interests most psychologists, it is primarily *people's* exchanges with each other and their surroundings that are relevant. To say that the genes contain a plan for developmental outcomes is to project our descriptions of the person backward in time to explain his or her coming into being.

Variation is fundamental to this story: Developmental systems are heterogeneous at every level, producing uniqueness and context specificity, and the social is integral to the construction of humans: not a finish coat applied to a biologically given natural object, but an aspect of the developmental complex, and involved in the very constitution of those biological object-subjects. Similarly, to say that subjects are interactively constructed is not to deny that some features may appear most of the time or even all of the time, although this often means we are simply not interested in the variations that exist. (Psychologists have more often been impressed by the regularity of language development than by linguistic differences among children, for instance.) A claim of universality will be convincing only in a given theoretical framework and at a given level of abstraction, and focus on such processes commits one neither to some vision of central control nor to the exclusion of experience.

Developmentalists have largely taken for granted the "pregivenness" of "the biological," even as they have disagreed about what features deserved the label. Common features, however, are no more dependent on the genes than variable ones are. And they are no less dependent on developmental environments, although those environments may be so predictable or stable that they recede into the analytic background unless insistently named. The context is not just a container, but part of a process in which it and the organism select and construct each other (Giddens 1987:98; Morris 1988; Oyama 1992; Venn 1984). We are ecologically embedded, social beings, although we are not always well served by our situations. It is entirely natural that our constructions of ourselves and of the world should be marked by our particular historical, cultural, social situations (although they will not necessarily "mirror" them), as well as whatever experiences may be common to those situations.

Shotter's (1984:42) descriptions of joint action, in which, without being aware of it, people alter their own and others' consciousnesses as they construct meanings together, may help us understand some of these developmental phenomena. In such joint action, whether antagonistic or cooperative, outcomes are often unpredictable, cannot be traced to the intentions of any single participant, and thus may seem to be external to them (p. 144). They can appear "natural" or "spontaneous," and it is easy to miss one's own (or others') contribution to them. This is a view of development as formation in a system composed of the person and the developmental environment, changing together. It is a view that does not permit nature-nurture distinctions, and that is what the natural-constructed distinction is.

The vision of mutually constraining influences that I present here does not, it should be evident by now, necessarily imply harmonious relations; indeed, I wish to avoid the implications of automatic stability, healthy function, or homeostatic regulation that are sometimes attached to the concept of system. One might find a vicious circle of interactions in a family, for example, that benefits no one very much but that is stable in the face of a variety of attempts to change it, thus fitting the usual idea of self-regulating systems. Similarly, ecological embeddedness does not entail benefit to all the organisms involved. Think of the costs of pregnancy to a mother, or the tight integration of the "traditional family" into existing political and economic structures (or the prey into any predator-prey system!).

RESPONSIBILITY

If one thinks of partially nested developmental systems that can be studied at a variety of levels; in which ongoing processes can be analyzed by provisionally designating some factors "causes" and others "effects," but in which causes and effects are not ultimately distinguishable; and in which organization need not be imposed on inert matter, but rather arises from matter in interaction (though not all organization is living organization: consider a crystal or a whirlpool), it may be possible to conceive responsibility in a different way. Rather than contrasting autonomously acting persons with passive objects, perhaps we can consider these to be two stances toward certain human interactions. One is oriented more toward considerations and consequences as seen from the agent's point(s) of view and occurring in a social context in which that agent is able to communicate acceptable, or at least intelligible, reasons for acting (responsible action, according to Shotter 1984:38, being a matter of shared interests, not of individuals). The other takes the point of view of some (third-person) observer. Moral agency does not require freedom from causes (what could this mean?) or even from biological causes. Rather, it requires, precisely, embeddedness in a causal world (see Dennett 1984). Only there can one be subject to the joys, pains, desires, and perplexities that give rise to action; only there can one affect the world; only there can one be *engaged* by the exchanges that constitute human life; only there can one be moved to encourage some outcomes and prevent others; and only there can one be positioned among others who regard one as responsible. Such positionings are not foregone, however. The earlier discussion of homosexuality shows how some people are attempting a strategic repositioning while others oppose it. In Shotter's "political economy of selfhood" (1984:117, 179), people enhance and limit each other's opportunities for development.

Immersion in the causal network, then, is not only consistent with moral action, it is the very condition for such action. An actor will frequently construe relevant antecedents and probable consequences rather differently from any given onlooker (including a psychologist). An important form of moral disputation, in fact, concerns the proper placement of action (its meaning) in life's colliding activities and understandings, influences and resultants. Blame and credit are also redistributed as people's understandings of their own actions change. It is possible, by emphasizing some connections over others, to allocate responsibility in a variety of ways, in what Kenneth Gergen (1990:587) calls the exercise of "practical rhetoric." Although rhetoric tends to have a degraded meaning today, persuasion and influence are hardly trivial matters. In disagreeing about responsibility, we deploy (partially) shared understandings, but we may also change those understandings. Included in these understandings are not only what are conventionally thought of as moral principles, but also more general beliefs about the world. Ideas of causality and possibility, for instance, so central to our developmental theories, are clearly relevant to these disputes: My intellectual and moral reasons for addressing them, finally, become difficult to separate.

A person is a unique node of confluences and divergences, a moving locus of interactions. Perhaps reasons can be seen as particular construals of one's present and potential locations in this network, in the light of particular considerations. They have to do with the meaning of acts. Responsible action (in the approbative sense of "action taken for reasons intelligible and acceptable to certain individuals") would then be action taken with respect to a suitably broad and detailed portion of the network, in a way that includes consideration of the impact of the act on an appropriate variety of others and on the inanimate world. To talk reasonably about agency and responsibility, that is, one should not ask whether people are subject to causal influences, because they always are. Rather, one should inquire what impact a person has on the world, and for what reasons.[6] Many ways of acting will enjoy some degree of consensual validation, but not all. Shaver (1985:vii) describes blaming as explaining by means of socially negotiated notions of negative consequences, causality, personal responsibility, and mitigation. Any of these is open to argument, as we saw in the discussion of homosexuality, so it is not possible to say what constitutes "suitably broad and detailed" deliberation *in general*. Because we each affect and are affected by others, in small ways and large, however, and because we are implicated in each other's very constitution as persons, the moral implications of our theories (and the theoretical implications of our morals), as well as the background of attitudes that informs them, deserve our most serious attention.

CONSTRUCTION OF THE NATURAL

In the view presented here, subjectivity *is* natural, and its development is an aspect of more general processes of construction. It is biological because it is an aspect of human life. This is radically to alter the usual meanings of *biological* and *natural*. No longer can they mark off a part of persons that comes ready-made. Rather than restricting the proper scope of biology, as many critics have wished to do, I broaden it to encompass the entire life cycle. Biology so construed cannot be used to parcel out developmental (or moral) responsibility to internal and external factors. Nor can it define an invariant core of human nature. Difficult issues of human capabilities—how to define the "could" in "could have done otherwise," for example—are not thereby resolved. I suspect that such issues, like accountability itself, are matters of complicated social negotiation, in which scientific understandings of possibility may play a part, but which are not definitively answerable by means of those understandings.

The developmental systems perspective is no less constructed than any other. Indeed, this book is part of its ongoing construction. This does not mean that it is arbitrary or indefensible, nor does it prevent me from believing passionately in it, or from hoping that if I explain adequately what considerations (scientific and otherwise) led me and others to formulate it, readers will come to see things as we do.

Naming Ourselves, Naming Others

We have looked at some of the ways our conceptions of development make it easier to see people as objects, formed and moved by causes, than as experiencing subjects who may act for reasons—that is, as persons.[7] These two attitudes carry moral implications, and the discussion of the politics and theory of homosexuality illustrated how they can be connected to conventional beliefs about biology and responsibility. I argued that we can see subjectivity as both biological and constructed, so that those who wish to speak of persons in all their particularity need not search for some nonbiological realm in which to do so. The life of the British mathematician Alan Turing invites us to contemplate our limited, but still important, power to name ourselves and others, to say what people are.

In his fascinating biography of this man, whose work on the concept of a "universal machine" was so important to the development of the modern computer, Andrew Hodges (1983) shows us someone who resisted being treated as an object to be manipulated and corrected. Turing was

perennially engaged in a struggle to reconcile body and mind, determinism and free will (or later, at least the appearance of free will [Hodges
1983:108]). In fact, the biography's two sections are called "The Physical" and "The Logical." These enduring concerns of Turing's, his central
role in initiating the cognitive revolution, and some aspects of his life and
death bring together the diverse topics I have touched on in this chapter.
Two perspectives on cognitive functioning found in Turing's writings, as
mindless mechanism and as mentality, articulate nicely with the questions of human subjectivity treated earlier.

STATES OF MIND AND INSTRUCTION NOTES

Turing marked two attitudes toward mental operations by naming them
differently (Hodges 1983:107–108). He spoke of a person, the original
"computer," doing computations as being at any moment in a "configuration" that was jointly determined by his own previous state and by
the symbol he had just read. Turing spoke of this configuration both as
a "state of mind" and as "instruction notes" that might stand in for a
state of mind (if, for instance, prior to interrupting his labors, the person
wrote a note to remind himself where he had left off). Although Turing
treated states and notes as interchangeable, and although these terms referred to the same configuration, one symbolized choice and will while
the other suggested the mindless execution of orders. These two ways of
thinking about cognitive activity appear to express the mathematician's
ambivalence about subjectivity and objectivity, agency and causation. It
is intriguing that although he forsook his earlier beliefs in immortal spirit
in favor of materialism, Turing allowed this ambivalence to remain in
his descriptions—both of the original human calculators and of the machines he devised to replace them. The uncertainty persists in contemporary science: We continue to depict humans sometimes as subjects with
consequential states of mind, and sometimes as mere consequences of
genetic or environmental input. This is why computers are such engaging
objects of theory and speculation. We wonder not just how like ourselves
they really are (Do they have minds? Do they think?), but also whether
we, like them, are "nothing but" machines (Is our experience of action
and choice mere illusion?).

To Turing, the computing machine promised to connect the world of abstract symbols with the physical world. Adding mathematical logic to existing notions of mind as machine (Hodges 1983:107) enabled an assemblage of tubes and wires to manipulate the very stuff, it seemed, of thought, and so to cross the Cartesian divide, joining mind and matter, reason and mechanism. His scheme is nevertheless a deterministic one, and as Turing seemed to realize, it is not useful to approach this divide by wondering whether people are determined or free, which is one of the ways the nature-nurture debate is employed. Venn remarks that only human beings, as God's chosen, rational creatures, can fill the void left by "the evacuation of the divine from matter" (1984:136). But what fills the gap when the reasoner him- or herself is left empty, when scientific explanation, couched in the language of genetic and environmental causes, leaves no room for human subjects? The space gapes insistently. As we saw, we are often impelled to fill it with quasi subjects. The gene, after all, has "reason" (information, programs, plans) and is Nature's chosen creature. Forbidden by science to be acting subjects, we must be objects — of the genetic (or even societal) "subjects" created in our own image.

Yet Turing's terms do not index two realities, one causal and one not, but rather two ways of approaching the same one. Although I do not find the image of a machine reading a symbol particularly fruitful in thinking about human experience, not least because it slights emotion, meaning, and context, it is convenient for my purposes that Turing placed determinative power in neither configuration nor symbol: They jointly determined the next configuration, and thus affected the impact of the next symbol. As I noted above, such joint determination eliminates the dichotomies of developmental control that bedevil the biological and social sciences alike, and simultaneously reinstates the person in the explanatory frame.[8] The difference between Turing's two attitudes is then seen to be a matter not of freedom versus causation, or activity versus passivity, but of point of view.

The language of genetic programming, in addition to drawing on older ideas of instinct, exploits the second of Turing's renditions of *configuration:* controlled by instructions. Indeed, the program is the conceptual descendant of these instructions, and it is useful to ask what role it plays in our understandings of development and behavior. Genetic programs are a way of expressing the implacable causality of "nature." They resemble instruction notes in connoting mindlessness. Like the ingenious mechanisms inside the automata said to have inspired Descartes, programs offer a way to explain apparently intelligent behavior without recourse to mentality. This use of *program* to refer to any control by a not-self (the "not-me in me" mentioned earlier) is interesting, for the first "instruction notes" were written by the human computer himself, to maintain continuity at a task. They were thus an expedient for *self* control. These instructions, however, have lent themselves to a quite different view of human activity, as controlled by something other than itself. Just as the instruction notes that made a "mindless executor" of Turing's human computer could either be stored inside the machine or supplied from the outside (Hodges 1983:302), behavior may be attributed to either genetic or environmental "information," or even to an outside brought inside (perhaps an introjected parent). In the end, what one wishes for is a view of causation and control that is more mobile, relational, and relativistic, in which investigations of local relationships are understood to depend on the observer's point of view and method of framing questions, in the way described in chapter 8.

TURING'S TEST

Turing, who was instrumental in breaking the Germans' secret codes during World War II, was a homosexual who underwent a course of chemical therapy after the war as an alternative to incarceration for "gross indecency." [9] He died not long afterward, apparently by suicide (Hodges 1983). The Turing test owes its name to this man. As it is understood today, the test states that any entity that can pass for human when communicating by means that conceal its physical makeup (perhaps by words displayed on a computer screen) is considered to have a mind, although this is not precisely the way Turing presented it. Turing (1964) himself called it the "imitation game," and its first players included a man, a woman, and an interrogator whose task it was to distinguish between them.

Turing's biographer comments, "Like any homosexual man, he was living an imitation game, not in the sense of conscious play-acting, but by being accepted as a person that he was not" (Hodges 1983:129). The author of the original imitation game did not himself play it very well, partly because he refused to accept the goal of being mistaken for a heterosexual. In homophobic postwar England, his downfall came from being all too forthcoming about his (illegal) homosexual self. When the estrogen he took as part of his "therapy" stimulated breast growth, he was apparently open about this as well.

The Turing test is a procedure that makes the body go away, leaving only an abstract pattern of pure, disembodied thought. It allows the entity being interrogated to communicate by a severely restricted channel, and to remain otherwise hidden. There is no room in this cognitivist story for feeling or desire, for messy materiality or morality (except, perhaps, insofar as these can be expressed symbolically). In fact, the abstract pattern's independence from its bodily instantiation is indispensable for the argument that machines can think.

Although one sometimes gets the idea that Turing wished to live an imitation game, "left alone in a room of his own, to deal with the outside world solely by rational argument" (Hodges 1983:425), in the end he did not, even to the extent of trying to hide his forbidden wants and acts. It is a distinct irony that this man, who struggled with the mind-body problem all his life and who in many ways minimized the embodied, social character of human mentality, should have been done in by such a crude vision of what bodies and minds—persons—are. In the end, he did not, of course, communicate with the world by pure intellect, and the self that was finally brought before the judicial system was not properly accountable there.

To Shotter (1984:147), to be autonomous is to be allowed to define the meaning of one's acts. That people cannot do this alone, however, is evident in his statement that this is possible only if one regulates oneself by socially shared criteria, so his distinction between expressing oneself and being defined by others is unclear. The considerations that brought Turing down were not shared by the whole society, and there were undoubtedly serious constraints on his ability to gain full personhood by naming himself (as responsible, etc.). But the power of others to name him, to define his self for him, was also limited, if only by his stubborn integrity and his obliviousness to aspects of his social world. Arguments about biological causation of the sort outlined earlier would not have saved Turing. The hormone treatment, in fact, was informed by assumptions about biological control of behavior. As the choice between imprisonment and hormones suggests, a coherent theory of causality was finally less crucial than the sheer unacceptability of public homosexuality (and, Hodges [1983:461] suggests, of certain breaches of class boundaries).

We began this chapter with a discussion of the Cartesian separation of mind from body that has influenced scientists' constructions of human subjectivity, as well as their understanding of other processes by which complex order arises through dynamic interaction. We saw that opposing genetic (bodily) causation to "the environment" or "culture" produces a peculiar mixture of causal and moral language. In this language, individuals may or may not be responsible for their contingent "nurture" (if they are not, their parents, societies, etc., must be, unless we ask whether the parents in turn are really responsible), but their genes are responsible for their fixed "nature." Many nature-nurture debates, in fact, have arisen because interested parties have accepted this logic and then battled to haul phenomena back and forth across the biological border. Rather than joining these tugs-of-war by maintaining that subjectivity is social, not natural, those who are concerned to reconstruct the subject must also reconstruct biology and development. In doing so, they will weaken the hold of the nature-nurture opposition on our science and our lives. The relations between this opposition and the one between body and mind are so many and so intimate that it is unlikely that we will make inroads on the troublesome aspects of Cartesian dualism without also confronting the ghost in the gene (and in society!) as well as the ghost in the machine.

I have tried to show that these are not only academic matters (a revealing locution) by turning to a politically fraught issue: the scientific, civil, and moral status of homosexuality. Without claiming to be able to foresee the consequences of some homosexuals' embracing a biological explanation for their erotic preferences, I expressed some misgivings about their decision to enter the nature-nurture fray. Turing's story provided background for the discussions of Cartesianism in cognitive science and the possibility of taking a variety of stances toward mentality. It also showed another limitation of some arguments over the causation of homosexuality: Not only are they often based on indefensible assumptions about developmental processes, they can be quite beside the point in a social order willing to correct its citizens to death.

Cushman (1991:218) asserts that "those who 'own' the self control our world" by prescribing healthy, proper behavior. By describing a person's coming into being, developmentalists characterize that person and that person's world; they are, as Cushman suggests, in a position of considerable influence. Scientific namings can also be moral namings.

Moral consideration includes evaluating an array of possible actions and consequences. The definition of this array is influenced by current scientific understandings, and surely a person's *actual* range of possible actions is partially constrained by current beliefs about that range. Too often, biological arguments, by virtue of their entanglement in a complex of beliefs about the fixity of nature, have prejudged these difficult matters. A reconstructed biology does not so readily lend itself to these uses. The responsible course would seem to require being as aware as possible of the myriad influences that inform our ways of being, acting, and knowing, and at the same time being alert to the ways in which what we do or say impinges on, informs, and even changes others. If we are to use biology to name, it should be a biology ample enough to include our whole selves as well as the social worlds in which we are made and which we help make—not in the disciplinary imperialism that sees ever broader compass for genetic control,[10] but rather in the attempt to reach far enough to describe (to paraphrase Shotter), what kind of world those genes "go on inside of."

11 Evolutionary and Developmental Formation:

Politics of the Boundary

Much of my work concerns the politics of the boundary. The meaning of *politics* here is very broad, having to do with all sorts of influence and power, but especially the power to define and privilege, include and exclude, render central or peripheral. Although this may involve matters "outside" science (a fraught frontier if ever there was one), it need not. Some of my reasons for working on the nature-nurture problem stem from concerns about publicly contested issues of, say, intelligence, race, or sex, but most have to do with the kinds of distinctions that are made in the scientific work that draws on and feeds these larger disputes.

Any theory carves the world in particular ways and so legitimates some entities and distinctions while leaving others beyond the pale — secondary, invisible, or unintelligible. Making the cellular or the nuclear membrane a primary theoretical boundary in the study of development and evolution, for example, may be justified by pointing to Weismann's barrier (1893), which in today's terms forbids "information flow" from the body to the DNA. This insistence on one-way movement of information is not just a matter of the presence or absence of cell-level DNA-altering feedback, however; it plays on and warrants the whole set of assumptions about inheritance and essence, permissible and impermissible explanations — even about academic disciplines — that we have been analyzing.

Inside or Outside Standard Evolutionary Theory?

I am aware that the words *politics* and *boundary* in my chapter title will raise some red flags. Part of my project is to inquire into the provocative power of such boundary-marking emblems. They tend to imply a particular field of oppositions, and they inevitably gloss over some differences and commonalities while proclaiming others, making it hard to move beyond global categorizations. This situation helps explain why there may be a need for something like a developmental systems perspective, to rework many of the basic oppositions in the biological and social sciences. It may also help us understand why developmental systems theory seems to be resisted for being simultaneously inside and outside existing theory: inside (and therefore redundant) because it stresses matters that are already acknowledged at least part of the time, and outside because it explicitly breaches certain sacred boundaries. A not uncommon reaction to DST is, "That's completely crazy, and besides, I already knew it."

Parts of this chapter will serve as a review of some of the basic arguments of the book. I start with the developmental systems case for broadening the concepts of inheritance and evolution, which means including within the hereditary package many factors that are traditionally considered to lie outside it. I also review some common defenses for the traditional positions. Next come some recent border disputes in evolutionary theory. In mentioning these disagreements I try to show that "science as usual" frequently involves disputes over what should be treated as internal and external to a given phenomenon. While the notion of a developmental system admittedly tampers with some categories that modern thinkers treat as sacrosanct, then, I'm suggesting that the kind of conceptual change it proposes is not different *in kind* from the everyday work of theorists, and that, indeed, boundary shifting is a favorite, although sometimes unacknowledged, technique in scientific dispute. This does not make it *right,* but it may make some of my own moves look less outlandish. Although DST is perhaps better known for questioning distinctions than for drawing them, repositioning a boundary can not only group together things that were formerly kept strictly apart, like genetic and other influences on development, it can also highlight new differences and bring the peripheral into focus. Trends in the direction of something like developmental systems are visible even in some standard theorizing.

At various points I touch on the selfish replicators that have figured in debates about "biological bases" of human behavior, and raise doubts about the status of "informational" units such as genes and "memes" as privileged currencies and explanatory tools. More inclusive conceptions of development and evolution open up space to address issues that have suffered relative neglect. It may even be, as I speculate in the penultimate section, that they will allow the exploration of the complexities of cooperation and competition without foreclosing some possibilities beforehand. I close with some comments on the position of DST with respect to more mainstream theories, suggesting that some of the resistance it has met stems from the fact that even in science, boundaries can be treated as defenses against heresy rather than as lines to be drawn and redrawn in the pursuit of various kinds of understanding. By speaking of the politics of biology in this way, I do not intend to expose the field to the world, unclothed and unlovely. The emperor has many clothes, more than he usually wears, and certain ensembles, not yet tried, might serve him better.

Redrawing Boundaries

THE BOUNDED GENE

The conceptual line drawn around the genes is fundamental to modern biology, and insofar as biological science plays a foundational role vis-à-vis other disciplines, other lines are drawn in accordance with it. It delimits inheritance proper, the genetic inheritance that was originally modeled on the inheritance of land, objects, or titles, but that now serves as the paradigm case. A semipermeable membrane around a certain kind of cell, or even around its nucleus, is thought to separate two kinds of "information," "transmitted" through different channels. Genetic information flows along the germ line; it passes through organisms without being (much) altered by the passage and continues for an indefinite number of generations. Hence the talk of potentially "immortal" genes (Williams 1966:24). The ontogenies of individual organisms are said to result from the decoding of this information, which may then be supplemented or modulated by the environment.

With a line drawn between the genes and everything else in the universe, it is perhaps not surprising that the channel for *non*genetic information remains underspecified (it may be cytoplasmic or extracellular, "experiential" or cultural), although a direct brain-to-brain route seems to be popular. Sober (1992) gives a taxonomy of selective models of cultural change. The models use either genes or learning as the mode of heredity, and either "having babies" or "having students" as the measure of fitness. We see here the centrality of information transmission as a way of accounting for continuity. Also evident is the intimate connection of these models with what I have been calling "developmental dualism," the doctrine that some developmental processes are (mostly) guided from within, and some are (mostly) guided from the outside. It becomes a simple matter to see how this conceptual geography encourages genetically reductive views of human affairs (if only gene replication counts), developmental dualism (especially if "memes," or cultural units, replicate too), or both. Neither is very pleasing.

The developmental systems approach allows us to redraw the overly restrictive boundary around the genes to include other developmentally important influences. The point is not to open a second channel, one to carry culture, but to trade discrete channels for interacting systems whose processes give rise to successive generations. Regularities in these systems' functioning support the sorts of predictable associations that make the language of transmission possible.

ASSUMING RELIABLE DEVELOPMENT

Transmission, whether of genes or of culture, is supposed to produce developmental regularity, but as I argued in chapter 5, it actually *presupposes* such regularity. When bodies, behavior patterns, or beliefs recur regularly, we can count on them without being able to explain them. When they don't, we may be at a loss to say why. How often do we accurately transmit even a single complex idea to a student or a colleague, and what else must be in place for this to be even thinkable? (And what would it mean to transmit a "single idea" in isolation?) Something is judged to have been transmitted when it reappears, through heaven knows what processes, in the "recipient." But invoking transmission not only fails to *explain* how this reconstruction occurs, it encourages us to ignore the fact that it must occur at all.

In a discussion of cultural evolution, Sober (1992) says that social scientists are interested in *causes* of differential transmission, and evolutionists in *consequences*. (Another version of this perspectival gap will come out when we talk about cooperation and competition below.) Thus Dawkins (1976:206) can blithely write of memes (his units of cultural evolution, analogous to genes) that "propagate themselves in the meme pool by leaping from brain to brain" precisely because, as Daniel Dennett (1995:359) acknowledges, evolutionists and cognitive scientists can "finesse" their "ignorance of the gory mechanical details of how the information got from A to B." One could suspect that more than finessing is involved here, given the evident glee with which these authors displace agency from humans to selfish replicators. Thus Dennett (1995:369, 471) speaks of "invasion of human brains by culture," and even of interacting "meme-infested brains" (not people!). True, in a more benign mood he says that brains are "furnished by" memes (1995:341), but his general thrust is conveyed by a description of selves as "created out of the interplay of memes exploiting and redirecting the machinery Mother Nature has given us" (1995:367). Actually, he *defines* the notion of a person by such memic invasion and manipulation.

These are especially vivid examples of the ways in which developmental issues and all sorts of other questions about social phenomena can be kept at bay (or worse, "explained") by evolutionary theorists. If evolutionists are content to *assume* the processes allowing reliable reappearance across generations, and wish only to document net outcomes, then one might reasonably ask what purpose is served by the agentic language, especially when it causes such dismay in some and lends itself to such misuse by others. Turning genes and memes into diminutive masterminds conflates process and consequence precisely in the way Patrick Bateson (1988b) warns against. Yet this sector of the literature continues to be populated by self-interested replicators tirelessly maneuvering for the reproductive edge. They are prototypes of competitive selfishness, and they design us to be their instruments, even, as discussed in chapter 9, providing our "real" motives and goals (see critiques in P. P. G. Bateson 1986; Kitcher 1985). Small wonder that this style of evolutionary writing has given rise to worries about the possibility, even the coherence, of cooperation, as well as about the more general implications of evolution for human lives.

PRIVILEGING GENES

In standard accounts of evolution and development, genes are privileged as both currency and cause, but the privileging is unjustified. Multileveled, changing contexts and processes can be reduced to the role of conduits (chapter 3) for information only by ignoring the "gory mechanical details" of the life cycle. Below are five typical rationales for gene-based accounts that have appeared in preceding chapters. Each is followed by a parenthesis indicating some of what is being glossed over. There is nothing sinister about these glossings over: People simplify all the time. Nor is there anything particularly heterodox about them. In fact, they are phrased in a conservative manner, as I think the kind of speaker I am imagining would qualify the arguments if pushed. Parentheses can, however, marginalize by typographical convention, minimizing what van der Weele (1999) calls the "conceptual room" allotted to a topic. She speaks of scientific choices in terms of the "ethics of attention." One can similarly speak, perhaps, of a *politics* of attention. So each parenthesis is then expanded, usually by parity of reasoning, or what philosopher Sidney Morganbesser (in Elster 1989:10) reportedly called the first law of Jewish logic: If *p*, why not *q?* The justification for giving special status to the genes is thus systematically (!) applied to other developmental influences as well. All of these influences and entities, both inside and outside the organism, are interactants in a developmental system that produces a life cycle. Herewith the arguments, with qualifying parentheses and their expansions:

I. ARGUMENT (to be read in a stentorian voice): Genes produce organisms.

QUALIFYING PARENTHESIS: (Although they are not, of course, sufficient; raw materials must be available and conditions must be adequate.)

DST EXPANSION OF PARENTHESIS: Genes themselves don't "make" anything, although they are involved in processes requiring many other molecules and conditions (Jablonka and Lamb 1995; Moss 1992; Neumann-Held 1999; Stent 1981; Strohman 1997). Other interactants (or resources, or means) are found at scales from the microscopic to the ecological, some living, some not. None is sufficient, and their effects are interdependent. Development never occurs (and could not occur) in a vacuum.

2. ARGUMENT: Shared genes are responsible for species characteristics.

QUALIFYING PARENTHESIS: (Again, as long as proper conditions are present.)

DST EXPANSION: Just as genes can't make organisms *in general,* they can't create species-typical characters *in particular.* Typical conditions, again at many scales, contribute to forming these characters, whose uniformity should not be exaggerated. The activity of the organism, including self-stimulation, is often a crucial aspect of species-typical development, and so are influences from other organisms (Gottlieb 1978, 1997; Johnston and Gottlieb 1990; Lehrman 1970; Oyama 1982). Genetic and environmental variation is often underestimated, and flexible processes can sometimes result in typical phenotypes despite atypical developmental resources.

3. ARGUMENT: Genetic variants specify the heritable phenotypes needed for natural selection.

QUALIFYING PARENTHESIS: (Of course, heritability depends on conditions, and it can be hard to separate genetic from environmental effects.)

DST EXPANSION: Unless nongenetic factors are excluded by stipulation, other developmental resources can also "specify" phenotypic variants, which can be heritable in a variety of senses. The genotype-phenotype correlations that warrant the talk of genetic specification may not occur under all circumstances, and may change within and across generational time. Specificity, furthermore, is a slippery matter; it depends on the question being asked, the comparison being made, and the measure being used, as well as the developmental state of the organisms and the context of the comparison. In fact, the genotype-environment correlations and statistical interactions that plague the behavior geneticist are manifestations of just the interdependent networks that developmental systems theorists describe (Nijhout 1990; K. C. Smith 1994).

4. ARGUMENT: Only genes are passed on in reproduction; phenotypes, and therefore environmental effects, are evanescent and thus evolutionarily irrelevant.

QUALIFYING PARENTHESIS: (Of course, the genes are housed in a cell.)

DST EXPANSION: If transmitting or "passing on" means "delivering materially unchanged," then few if any developmental resources are transmitted across evolutionary time, depending on how one measures material change. If transmission means "reliably present in the next life cycle," which is the biologically relevant meaning in DST, then an indefinitely large set of heterogeneous resources or means is transmitted. They are sought or produced by the organism itself, supplied by other organisms, perhaps through social processes and institutions, or are otherwise available (Caporael 1997; Ingold 1995). Although many developmentally important environmental features are exceedingly stable, others are non-continuous, perhaps varying seasonally or geographically (Griffiths and Gray 1994). Any definition of inheritance that doesn't privilege the nuclear or cell boundary a priori will be applicable to other constituents of the system: If *p*, why not *q?* The developmental systems perspective stresses the processes that bring together the prerequisites for successive iterations of a life cycle.

As I observed in chapter 4, bits of DNA could magically materialize whenever they were to be used: It is appearance at the right time and place that counts, not material continuity. The more we learn about how cells carry on, the less fanciful this idea seems; a functional gene may not consist of a continuous stretch of DNA but may have to be assembled on the spot by quite elaborate methods (Ho 1988b; Moss 1992; Neumann-Held 1999).

5. ARGUMENT: If gene frequencies don't change, then evolution has not, by definition, occurred.

QUALIFYING PARENTHESIS: (Of course, the gene concept is relatively recent, and other definitions are possible.)

DST EXPANSION: A historian could tell us how gene frequencies moved from being an *index* of evolutionary change to being *definitional* (see papers in Keller and Lloyd 1992), but we needn't insist on that one definition. In fact, many branches of biology routinely speak of changes in phenotypes (Brandon 1990:5; Johnston and Gottlieb 1990; and, in a different spirit, Maynard Smith 1984, on game theory). If one must have a "unit" of evolution, it would be the interactive developmental system:

life cycles of organisms in their niches. Evolution would then be change in the constitution and distribution of these systems. This definition embraces, but is not restricted to, more traditional ones.

Negotiations at the Border

The preceding list should give a sense of DST's approach to development and evolution. Its proponents extend some boundaries, but in many ways are simply making explicit what is everywhere implicit, and what increasing numbers of workers are saying more or less directly. In another, more realistic, sense, however, it looks like an uphill fight all the way. That these are both true is in itself significant, for it reveals the complexity of the process of theoretical conservatism and change (Shanahan 1997; B. H. Smith 1997).

Robert Brandon and Janis Antonovics (1995:229), for instance, claim that organisms and their environments coevolve, and discuss Lewontin's (1983b) argument that organisms and environments are interconnected in ways that require a reconceptualization of evolution. They note that the idea of mutual influence has been around for a long time—in the practice of rotating crops, for example. It is found in the evolutionary literature, too, often in asides and ceteris paribus clauses, but prominently in game theory, in which competing organisms take each other's behavior into account.

In his comments on Brandon and Antonovics's article, Wolters (1995) is appreciative of their proposal but disagrees with them on whether conspecifics are internal or external to the population. Border disputes like these are everyday affairs. Such uncertainties, like the long-standing one about whether the niche belongs to organisms or to the environment, illustrate both the perpetual ambiguity of these divides and scientists' need to bridge them.

USING WHAT YOU "KNOW"

Theorists are annoyed when they are told what they have "always known." Yet there is a difference between knowing in a parenthetical, "of course it's important" way about the intimacy and reciprocality of organism-environment exchanges in development and evolution, say, and incorporating the knowledge into models and explanations, research and theory. Indeed, one of Brandon and Antonovics's stated goals is to redress evolutionary biology's failure to integrate this "knowledge" of interdependence. Van der Weele (1999), remarking on the same phenomenon in developmental biology, shows some of the ways the role of the environment is marginalized. Because any dividing line exists on a complicated, multidimensional landscape of belief and practice, it can implicate all sorts of other affirmations and denials, to say nothing of loyalties and betrayals. A great deal of energy is devoted to managing them.

MOVING THE LINE

Negotiations over entities and explanations are integral to the making of science, including the continuous "making" of science by marking the borders between it and nonscience. Whenever the scope of a theory is being explored or unexpected results must be accommodated, limits are put into question. A common response to theoretical challenge is to engage in strategic boundary work. Hull (1988:202), for instance, holds that the proponents of the evolutionary synthesis simply redefined the theory to include neutralism.[1]

As I suggested earlier, one can also attempt to place an opponent out-of-bounds, perhaps by calling him or her a Lamarckian. There is an interesting story to be told about this use of Lamarck against deviations from traditional conceptions of development and evolution, but I will not tell it here (see Hull 1988:chap. 12; Jablonka and Lamb 1995; Sterelny, Smith, and Dickison 1996). Suffice it to say that such name-calling is often a dubious attempt to reinforce dubious distinctions between nature and nurture, or to distinguish between hardheaded, scientific neo-Darwinism and its fuzzyheaded, sentimental detractors. Conventional wisdom may, with some imprecision, call cultural change "Lamarckian evolution," but it seems that biology itself must be shielded by Weismann's barrier or some other line between the genetic and the environmental. This is the case despite the difficulty of saying just where the line lies and despite the myriad connections and interdependencies between the divided territories (Latour 1993).

We have seen that variously located membranes are used simultaneously to demarcate kinds of information, inheritance, transmission, evolution, even science itself. Most of the lines touched on here have been, as they say, internal to science, but they are political nonetheless in the general sense mentioned earlier: aspects of disciplinary and doctrinal influence. The skeins of citations are also boundary work, as are the lacunae left when critics are answered or dismissed without being named. Thus are ancestors and bastards created, in-groups and out-groups, in work that is *at once* intellectual and political.

The runoff from these controversies is seldom contained by ivy-covered walls. Debates about macroevolution or developmental constraints, for instance, have been recruited in the battle between creation science and orthodox science. Part of the emotional charge to these debates comes from exasperation that a battle considered long won is not over after all. Theological considerations were once standard in biological inference. That changed, and now some would change things back; theorists accustomed to talking about escalating arms races should expect to face updated weapons. As Hull (1988:73) remarks, "The dispute over Darwinism was as much a disagreement over the nature of genuine science as over the existence of evolution." It still is: Questioning received evolutionary wisdom these days can open one to charges of crypto-creationism. This can be a potent way of keeping dissenters in line.

Replicators, Cooperation, and Competition

Once one has accepted a gene-based view of biological evolution one can then ask: If genes evolve in particular ways, why not other things? If tiny particles of matter are central to one story, atoms of thought or culture could populate another (note once again the body-mind duality). Many approaches to learning, society, and culture have been inspired by evolutionary theory. Sometimes social arrangements are explained by biology. Or the relation between biological evolution and cultural evolution may be analogical: They may "interact" or compete, or run in parallel. We might then envision gene-meme (or culture) coevolution. Just as genes could manipulate a bird into making a nest to ensure their own propagation (Dawkins 1982), an idea or tune could arrange for its own replication by imitation (hence the term *replicator*). There is some parity of reasoning going on here, and scheming memes can be as entertaining as selfish genes. But as Patrick Bateson (1978) suggests in a teasing but pointed rejoinder to Dawkins, one could also speak of nests using genes to make more nests.

A PROLIFERATION OF REPLICATORS

Serious proposals to enlarge the evolutionary cast of characters tend to meet resistance. One of the most conservative elements in standard views is a belief in entities whose "informational" nature gives them special significance in both evolutionary and developmental formation. "Conservative" here refers to theory, but one of the reasons evolutionary theory commands so much attention from scientists and nonscientists alike is its entanglement in larger political and moral questions. Having to announce "I am not a Nazi" or "I am not a lackey of the capitalist power structure" is about as much fun as going on television to say to the people who elected you, "I am not a crook." [2] Scientists offended at having to parry such accusations may, however, have helped prepare the way for them by cleaving to a conceptual scheme in which human relations are explained by the competition of immortal quasi agents that get themselves counted in the next generation by making and running the bodies that carry them; or, less melodramatically, a scheme in which a biologically conservative "nature" constrains and limits "nurture," defining beforehand the range of possible variation.

The lively and ongoing exchange between Kim Sterelny, Kelly Smith, and Michael Dickison (1996) and Paul Griffiths and Russell Gray (1994, 1997) is, among other things, a second-generation Dawkins-Bateson disagreement about the notion of the selfish replicator. For Dawkins 1982 (Hull 1988 has a different view), a replicator is an entity that makes organisms and is copied at differing rates, depending on the organisms' reproductive success (see Griesemer in press on copying). Sterelny, Smith, and Dickison insist on the privileged *evolutionary* status of the replicator, although they accept "the radicals'" critique of its special role in controlling *development*. (For "radicals" read "DST.") Dawkins and many others see this privileged role in developmental causation in the guise of genetic instructions, programs, and the like. The "reformists" Sterelny, Smith, and Dickison (my term, not theirs) not only wish to *retain* evolutionary replicators, they want to enlarge the category to include more entities: "a still more raucous and motley crowd of squabbling replicators," like nests and burrows, which can increase their chances of appearing in the next generation by influencing their inhabitants (1996:400). These authors note that treating such objects as replicators is consistent with Dawkins's "basic conceptual structure," if not with his "actual practice" (see also Gray 1992). But DST-ers Griffiths and Gray question the very idea of privileged replicators. The choice, to simplify somewhat, is between re-forming the game by admitting more players, and changing it. This game is, of course, enclosed in a bigger game that is not being questioned, maybe the game of "evolutionary theory" or "scholarly exchange." [3]

The debate is too complex to summarize here, but it is interesting partly because evolutionary theory, which eliminated divine agency from its explanation of life, may end up postulating surrogate agents. This metaphorical talk, discussed in chapter 10, is not entirely benign, and the difficulties are not entirely deflected by accusing critics of being squeam-ish about real life; nor are they completely defused by invoking poetic license. For one thing, such talk makes the already difficult task of think-ing about human behavior even harder. In most cases this is probably an unintended consequence.[4] Yet, selfish-gene talk often seems an aspect of a more general evolutionary machismo directed against anyone foolish enough to think that nature (or humanity) is nice.

Although I don't think nature is particularly nice, people sometimes are, and their niceness is not always a selfish strategy. I am concerned about the kinds of truths evolution is thought to render, and I believe that the problems raised by current orthodoxy's genecentrism go beyond the theoretical issues sketched earlier. The risk that we will see humans as being driven by the self-interested replicators that "infest" their bodies and brains is somewhat diluted if burrows and nests can be replicators, too. But crowds of contentiously quarreling quasi agents may not be the best basis for an adequate view of human life. We are left with more com-petitors, not a different view of competition.

EVOLVING SYSTEMS

If one thinks of evolution as change in the constitution and distribution of developmental systems, it is certainly reasonable to record the reappearance in successive generations of particular features, variant or otherwise, as a way of tracking the characteristics of life cycles. Historical considerations, ease of measurement, and amenability to modeling will inform the choice of features and comparisons. Stability and change in associations among features can also be of interest: This is what studies of heritability, linkage, habitat imprinting, social structure, and migration are about. Virtually all the familiar kinds of evolutionary research can still be done against such a broadened background, but there are no "informationally" or causally privileged elements to drive the entire process. There is, though, an emphasis on the many explicit and tacit choices made by the analyst who temporarily focuses on certain elements as "informative." In addition, areas to some extent excluded from the synthetic theory of evolution, such as development, biogeography, and ecology (Gray 1988), are comfortably accommodated by DST.

Brandon and Antonovics (1995) argue that the kinds of complex relationships they document have been practically invisible until now. I submit that thinking in terms of evolving systems rather than disembodied gene pools or genetic programs makes such phenomena salient. Brandon and Antonovics's term, "coevolution," though, suggests the linked change in two distinct units, brought about by natural selection of each by the other (typically different species, as in the host-parasite models the authors say are closest to their own, 1995:227). I am not arguing for any single definition of coevolution (see Nitecki 1983 for a sampling of the possibilities). I do suspect that, to the extent that Brandon and Antonovics treat their plants and environments as somehow analogous to two species, they fall short of effectively serving their stated goal of implementing Lewontin's insights into organism-environment interpenetration. To the extent that they approach that goal, they also approach something that looks more and more like a developmental system.

Griffiths and Gray (1997) compare the organism-environment coevolution of Brandon and Antonovics (1995) with organism-environment relations in evolving developmental systems, taking up some complaints that have been raised about the latter. One is that such systems supposedly involve an intractable degree of complexity. Griffiths and Gray, though, remind us that any *actual* investigation is necessarily limited in scope. They remark on the gap between the rhetorical force of a *charge* of wooly holism (leveled by Sterelny, Smith, and Dickison 1996) and its *practical significance,* pointing out that a "complete" picture of Brandon and Antonovics's organism-environment coevolution would be as complex as a "complete" developmental systems account. Any real analysis, though, is circumscribed. (On the goal of completeness, see van der Weele 1999.)

Earlier I speculated about sources of resistance to the pleasing (to me) unity and inclusiveness of the developmental systems account. These sources included the fear of the Lamarckian heresy. But one need not haul out poor Lamarck in order to complain that speaking of evolving organism-environment systems involves an unacceptably great change from existing usage. Even Brandon and Antonovics's largely sympathetic commentator (Wolters 1995) feels that it's a bit much to talk about the *environment* evolving.[5] Apparently defining evolution by selection, Wolters also seems to say that environments are not naturally selected, so they *change* but do not *evolve.*

But Gray (1988) points out that Darwinism has no timeless essence. "Existing usage" is seldom univocal; common terms can hide decidedly varied meanings, and over the years theories can be marvelously elastic.[6] At any moment, the amount of heterogeneity of opinion within a tradition can be stressed or minimized; when threatened from "the outside" we circle our wagons and face outward rather than detailing our internal disputes. Hull (1988:200), for one, opines that "in retrospect, Huxley's appellation ["Synthetic Theory"] was hardly based on past accomplishments but was a combination of a public-relations ploy and a hope for the future." Keller and Lloyd's (1992) *Keywords in Evolutionary Biology* shows that even fundamental concepts such as gene, fitness, selection, and species are anything but monolithic. This does not keep them from being used, and it doesn't necessarily hinder discussion. Sometimes heterogeneity can feed flexibility and extension (Fujimura 1992). Revisionist developmental systems workers (dis)respectfully submit additional concepts, such as information, transmission, inheritance, and even evolution itself, for scrutiny and reevaluation.

INDICATORS AND INTERESTS

Some worry that humans as a species are necessarily warlike, or patriar-chal. Others offer counterclaims that we are basically peaceful, or matri-archal. As we saw in chapter 7, these dynamics also occur with respect to intraspecific differences. To speak of biology is often to speak of what is deep and really real, so it is not surprising that people feel the need to fight evolutionary fire with more (albeit nicer) evolutionary fire, perhaps countering innate selfishness with innate altruism (Kohn 1990; comments in Oyama 1989). This is unnecessary, and in the long run counterproduc-tive. Nor is the solution to say that we have no nature, for this is apt to be taken as a claim that we can be *anything,* or else as a denial of within-species commonalities. To avoid these traps, I have recommended that *nature* simply refer to the organism's characteristics. Natures change, and organisms with the same genotype can have different natures. *Nurture* then refers to the developmental processes that make and change these natures.

Earlier I alluded to the increasingly widespread conviction that self-ishness is part of our biological nature (in the usual senses, as basic or inescapable). It has become hard to think of organisms, including our-selves, as anything but competitive. This is not only due to images of striving replicators. Assuming cooperation to be, at best, a competitive strategy can make it conceptually unstable. Keller (1992) observes that while mutualisms and cooperative relationships are acknowledged in bi-ology, they tend to be subordinated to competitive relationships, at times even being called "cooperative competition." Competition is seen to be *caused* by limited resources or is simply *defined* by them. Resources are assumed to be measurable without reference to organisms, Keller re-ports, and an organism's consumption is similarly treated independently of other organisms. All this contributes to the virtual unthinkability of cooperative relationships except as they are subservient to competition. Keller further remarks on biologists' apparent inability to retain certain periodically rediscovered insights about the dynamics of cooperation. This looks like insufficient conceptual room, or, as she puts it, "air space" (1992:120).

The biological literature is full of strategic altruism and mutual back-scratching, but always with an eye to reproductive advantage. Why must we keep *our* eyes on reproductive advantage, itself a multivocal notion? Because, we're told, natural selection's invisible hand is guided by an eye fixed on the bottom line. But why just one bottom line?

Keller (1992) reports that evolutionary theorists speak of competition in the absence of direct conflict or contact among the organisms in question, even when they are not using the same resource: The relation is competitive because the *analyst* is making the comparison. "This extension, where 'competition' can cover all possible circumstances of relative viability and reproductivity, brings with it, then, the tendency to equate competition with natural selection itself" (p. 125). To continue the conceptual slide, remember that some people virtually identify evolution with natural selection, which is typically defined by genetic change. (Recall the comment on Wolters, above.) Peter Taylor (pers. comm. 1996) confirms that researchers often do not actually *observe* the use of a limited resource by different species, and that multiple indirect effects confound the inverse relations between population sizes from which ecologists have typically inferred competition.

Let's reflect for a moment on genetic advantage. Dennett (1995:327–329) defends the idea that genes can have interests by comparing them to any other entity for whose benefit things are done, like children, corporations, and ideas. Natural selection is directed at advancing the welfare of genes, he says, so genes have interests. Although he stops a hair short of flatly denying that all this is basically bookkeeping (Wimsatt 1984), he makes it quite clear that he thinks these "interests" are not just a matter of what the biologist is counting. One gets the impression that interests are conferred by Dame Selection, not by the scientist.

One can certainly track allele (genetic variant) frequencies from generation to generation, *if this is what one wants to know.* Similarly, one can count the noses of a motley of replicators, but one should not forget who is choosing to count, and one should avoid straitening the conceptual space by adopting overly restrictive definitions or by confusing the interests of the counter with those of the countee. The dominance of population genetic definitions was established with the evolutionary synthesis, which has its share of gaps, loose ends, and indeterminacies. Without being so audacious as to present "better" definitions of cooperation and competition, I want to speculate a bit about collective activity.

Developmental Systems–Style Competition and Cooperation

"The radicals" should not be cast as exclusive press agents for cooperation. As I noted above, evolutionists have studied it intensively, and Griffiths and Gray (1994) stress the importance of competition (as dependence on limited shared resources) in DST. The idea of competition is considerably altered, though, when the focus shifts from organisms or agentlike replicators to systems of interrelated processes (see Griffiths and Gray 1997). At issue is the evolutionary *role* given to the complex interdependencies and integration emphasized in DST. Below are some tentative comments about these ideas in what might loosely be called a developmental systems style: one that moves with a certain fluidity among scales and measures, taking a pragmatic stance toward research decisions and a somewhat skeptical one toward much received wisdom. This approach emphasizes emergent pattern from shifting, heterogeneous sets of interactants, and changing, multiple control (see also Taylor 1995). Such phenomena, which some call "cooperative," are frequently described by ecologists, developmentalists, and social scientists. They tend to appear in evolutionary theory only as constraints on, or conditions for, selection — as messy details to be bracketed out, or as strategies to propagate genes. Although present in the theory, then, they are rendered conceptually secondary. It is possible to avoid that kind of prior privileging. These aspects of biological processes can be treated in a way that does not always measure them against the same bottom line. One can, for instance, concentrate on consequences of variant developmental systems' differing rates of self-perpetuation or on the interrelated causes involved in the systems themselves.[7] In the latter case, the "bottom line" is the continuity of a system's functioning, whether it is reliably repeated or not.

In chapter 6 I spoke of the importance of understanding how a set of processes *keeps going* partly by re-creating its own constituents (N.B.: *not* how some subset of constituents re-creates *itself*, making and using the rest of the system to further its own proliferative interests). I was speaking of autocatalytic chemical reactions in vital processes, but the significance of interacting, self-perpetuating, and self-preparing complexes of resources and processes is more general. Larger-scale interactions are implicated in reproduction and ontogeny as well, and they are not necessarily contained within a skin (this is one of the points of Dawkins's 1982 *Extended Phenotype,* and what made that exploration so interesting to me). The continuity of the germ line is supposed to explain the repeated cycling of life courses, but this leaves out the rest of the developmental complex, downgrading important phenomena and encouraging problematic conceptions of all-knowing designer genes. Disclaimers of complete genetic efficacy are becoming more common, but they tend to leave the basic message of powerful and clever macromolecules largely untouched. If there is indeed a practical difference between merely acknowledging phenomena and giving them full-fledged theoretical status, then it could be worthwhile to look at evolution through different lenses.

If we ask how systems keep going, how they change or remain stable, how changes at one level are, or are not, reflected at others, there will be room for tracking particular units from generation to generation. They will not have unique status a priori, so analysts may be called on to justify their decisions. A benefit of refocusing from tightly bounded self-replicators to loosely bounded repeating systems, however, might be to allow a more discriminating and generous view of collective activity (Gordon in press). Just as development has never been fully integrated into the synthetic theory, the question of long-term stability of multi-species communities has not been adequately dealt with either (Taylor 1992; see also Gray 1988). Peter Taylor stresses the historicity of ecological complexes, which must be developed or reconstructed, not just "dispersed."

Faced with a complex system made of processes at scales from the molecular to the biogeographical, it is necessary to ask: Cooperation for *what?* Competition for *what?* There need not be just one bottom line (B. H. Smith 1988:chap. 6), and for any indicator to track *anything* the system *must be kept going*. Processes and entities at diverse scales are often at least partially nested, and there may be different consequences of an interaction at different levels (Wimsatt 1984). Hence my earlier allusion to competing theorists "cooperating" to sustain larger games, which may themselves be in higher-level competition. At any level, the entities or processes must be scrupulously specified. The same is true of the relevant outcomes. If an analyst invokes competition when comparing outcomes (recall Keller's point), the counterpart could be to speak of cooperation any time an interaction contributes to the creation or maintenance of the outcomes themselves, though using loaded terms such as *competition* and *cooperation* for such broad categories of consequences and causes is probably ill-advised. In any case, these seem to be different kinds of questions, such that either can be subsumed by the other by embedding. If the highest-level game that is being monitored is about relative *representation* in an outcome, then "cooperative competition" could make sense. If it is about *contributions* to an outcome, then "competitive cooperation" becomes intelligible. For both kinds of questions, a variety of indicators and currencies could be considered, but one need not be automatically subordinated to the other; which is primary depends on the analytic question. It would then become difficult to conclude anything about the "basic" nature of behavior—or of life in general.

Taking this kind of multilayered approach to human interactions reveals social behavior whose variety, shifting subtlety, and complexity defy straightforward categorization. Think of Goffman's (1959) classic social-psychological analysis of self-presentation. People manage these presentations in ways that may be viewed as competitive. At the same time, they enter into wordless collusion to preserve a certain definition of the situation. The mobile psychology of "us" and "them" can also be of interest here (Caporael et al. 1989); think of the people/things you might be proud of, defensive of, embarrassed for, and the conditions under which those propensities can be aroused and altered, sometimes moment-to-moment. Trust is also crucial in our collective action, including science (P. Bateson 1988b; Shapin 1994).

Locating Developmental Systems

I am not advocating that we abandon existing methods of studying competition and cooperation, but rather that we take very seriously their limitations, and then ask, Is there anything else, perhaps quite different, that we might want to know, as scientists or as citizens? The shape of the theoretical background surely influences the ease with which, and the ways in which, other questions come to mind, as well as our interpretive resources.

Newspapers tell us that a single product can come to dominate a market for reasons other than absolute superiority. The essays in *Keywords in Evolutionary Biology* (Keller and Lloyd 1992) are instructive, not only for the contemporaneous and historical variety of definitions, methods, and assumptions surrounding important concepts, but for the stories that can be read between the lines and between the entries. These are stories about the dynamics of a sometimes loosely associated set of developing research areas, impinging on each other, requiring coordinating concepts, flourishing or not (Callon 1991; Taylor 1995). A historical perspective can also help us understand how such topics as cooperation and environmental influences in evolution (other than selective ones) gain the power to contaminate, in these cases by association with group selection and Lamarck (Shanahan 1997).

When I speak of a politics of theoretical boundaries, I do not mean that people are always maneuvering to push some views and suppress others, although they often are. But unintended consequences are ubiquitous. There is a whole shadow domain of events and connections that are hard to relate to anybody's intent (Shotter 1984). Convenient measurements tend to turn into definitions, and not only in science, despite individuals' explicit desires that this not happen. (Think of IQ, or grading in school. Psychologists are taught that Alfred Binet wanted to avoid precisely the use to which intelligence tests, some of which still bear his name, were eventually put.) Specialized measuring technologies arise, squeezing out other perspectives as intellectual and professional interests become linked in complex networks. Certain kinds of analyses and explanations are readily available, while others are hard even to think about. A colleague confides, "We don't have the language" to describe social behavior.

Developmental systems treatments have sometimes been criticized for being too radical. Yet when faced with DST's specific points, people often reply that the matter in question has always been acknowledged within conventional theory. I would not want the potential usefulness of this approach to be missed because topics such as cooperation (not assessed in an exclusively competitive frame) and environmental influences (not treated exclusively from the point of view of natural selection) have been stigmatized. Classification as "radical" or "moderate" implies a single point from which distance can be measured, but I don't think that contemporary evolutionary theory has such a center. Perhaps natural selection comes close. Nor would I want what I believe are valid criticisms to be hastily brushed off with a reflex, "We knew that," when what is needed is something more. Some of what is said about developmental systems is indeed acknowledged by most evolutionists, but one doesn't have to be a developmental systems theorist to see the difference between some familiar-looking bits and a novel but consistent configuration. Clarifying the relationships between this perspective and others (I consider several in Oyama 1999; see also van der Weele 1999) may help bypass some reflex objections and answer others, but the judgment about whether work on developmental systems lies inside or outside the synthetic theory can probably be made only retrospectively. Even then there is no guarantee that the judgment will be either unanimous or enduring.

Notes

Introduction

1 I use DST to refer to a set of concepts, methods, and reformulations of certain fundamental ideas in biology and the social sciences. The term should not be taken to imply a congealed party line or to have other connotations of "theory" more narrowly construed. In chapter 11, I discuss some of the considerations involved in delineating the changing groups of people and ideas that make up such alternative approaches.

   Developmental system has also been used by others for different complexes of ideas, usually in a less formal manner (but see note 15). As always, there is the possibility of confusion, but this book should make clear when a coincidence of terms is just that, and when it signals substantive kinship.

2 Some of this can be found elsewhere. See van der Weele's (1999) comparison of developmental systems theory with process structuralism and neo-Darwinism, and my (1999) elaboration and extension of her treatment to include the autopoeisis of Maturana and Varela. There are also some brief comments on Gibsonian ecological psychology in Oyama (1990). See Godfrey-Smith (1996) as well. Fuller treatment of some of these matters appears in Oyama, Griffiths and Gray (in press) and in the reissue of my *Ontogeny of Information* (2000).

3 Gray (1992) gives a nice description of the basic tenets of what he calls "constructionism" (what I here term "constructivist interactionism"), Griffiths and Gray (1994, 1997) deploy many of them to address some fundamental issues in evolutionary biology, and Griffiths (1997) does the same for the emotions.

   For works on developmental biology using a similar view of development, see Nijhout (1990) and van der Weele (1999), and to see what it looks like in molecular biology and genetics, see Moss (1992), Neumann-Held (1999), and K. C. Smith (1994). The dynamical systems approach, which has a great deal in common with the one under discussion, is employed by Fogel (1993) and Thelen and Smith (1994) in developmental psychology, and by Hendriks-Jansen (1996) and van Gelder and Port (1995) in cognitive science. For a different use of dynamical systems that nevertheless articulates in interesting ways with some of the ideas explored in this book, see Fontana and Buss (1996). Caporael (1997) gives an account of human social psychological evolution that takes issue with certain of the trends in contemporary evolutionary psychology, and Andresen (1992) and Studdert-Kennedy (1990) have written on language in ways that are conceptually related to the views being presented here.

4   There is no antipathy to genes here, although one does need to say just what sense of *gene* one intends. There is, however, a dissatisfaction with certain ways of dealing with them. The diversity of meanings of *gene* contributes to its rhetorical power. See Keller and Lloyd (1992) for some discussions. Griffiths and Neumann-Held (1999) provide alternative understandings of the "molecular gene" and the "evolutionary gene." They argue that the evolutionists' difference makers need not correspond to specific stretches of DNA, and show some of the implications of this idea.

5   Slow, that is, if we take our own quarter-century chunks as paradigmatic. Other organisms cycle more quickly or slowly, which can make for interesting and sometimes consequential dynamics. Consider the ways in which our attempts to deal with disease or insect pests can encourage their quite rapid evolution into ever more formidable threats, in what evolutionists sometimes call "arms races." Diseases and pests, in turn, are important influences on human evolution.

   The conscious changes of scale mentioned here, incidentally, are to be distinguished from the unacknowledged ones that sometimes result in the bizarre confusions of genotypes with phenotypes that crop up in the literature.

6   This is so whether the subject is the ontogeny of organisms or the evolutionary relationship between development and natural selection. I have in mind here people like Alberch (1982a); Gould & Lewontin (1979); Ho and Saunders (1979); Maynard Smith, Burian, Kauffman, Alberch, Campbell, Goodwin, Lande, Raup, and Wolpert (1985); and Webster and Goodwin (1982). See discussions in Amundson (1994); K. C. Smith (1994); Oyama, 1992, 1993; and van der Weele (1993).

   S. J. Gould (1992, pp. 276–277) observes that "nothing in the strict Darwinian paradigm suggests hostility to developmental issues, but the theory offers precious little space for the major internalist and structuralist themes of embryology." This quote illustrates the view of development as structuralist *inside* to natural selection's *outside*. At the same time, it points out the difference between a de jure theoretical entailment and a de facto practice. That is, consideration of development is not actually ruled out of evolutionary theory. It is just because there is no absolute theoretical exclusion, even while there is some degree of exclusion in practice, that it is difficult to raise objections about the current state of affairs.

7   "Process structuralists" such as Goodwin (1989, p. 96) and Ho and Saunders (1979) have criticized the gene-frequency definition as well.

8   Gilbert, Opitz, and Raff (1996, p. 358) date this definition from 1937. Those advocating a broader construal are now in a distinct minority. The definition I have used is "change in constitution and distribution of developmental systems"; Griffiths and Gray (1994) offer another.

9   Ho (1984) and Jablonka & Lamb (1995) are more ample accounts, detailing a multitude of phenomena that are missed by an exclusive focus on genes, even without leaving the molecular level.

10 The organism, poor thing. It is a dead end because reproduction is seen as a process that passes on genes; the soma or body is left behind, along with any record it may bear of its own life history. Lewontin (1992, pp. 137–138) calls this Mendelism's "causal rupture" between inheritance and development. For some history of genetic terminology, as well as many more topics relevant to the problems taken up in *Evolution's Eye,* see papers in Keller and Lloyd (1992).

11 Gene heaven is, of course, an unending succession of gene pools—witness some evolutionary theorists' invocation of genetic immortality. Compare, too, the differing fates of the body and the soul with those of the phenotype and the genotype.

12 Nelkin and Lindee's (1995) book on the gene as icon gives many examples, along with some history and many references to the popular and scientific literature.

13 Over the last several decades such dichotomies and their interrelations have received considerable attention in cultural studies, philosophy of science, and associated fields. For discussion, see Haraway (1989, 1991); Latour (1987, 1993); and B. H. Smith & Plotnitsky (1997).

14 This is also a reason to refer to my own approach as "constructivist interactionism"—that is, to distinguish it from this kind of social constructionism. There has been little consistency in terminology in this area, and every choice has its problems, some of which I discuss in chapter 3.

15 This is in contrast to Ford and Lerner's (1992) treatment, even though they call their approach developmental systems theory and consider it compatible with my own (p. 208). For them, perception seems to give access to already existing information in the world, whereas for me, information has no fixed location. Rather, it marks a particular distinction being made from a particular vantage point; see Oyama (2000).

1 Transmission and Construction: Levels and the Problem of Heredity

This was based on a talk given at the 1985 Schneirla Conference. It appears in slightly different form in G. Greenberg and E. Tobach (Eds.) (1992), *Levels of social behavior: Evolutionary and genetic aspects* (pp. 51–60). Wichita, KA: T. C. Schneirla Research Fund.

1 By using the word *wish* I do not mean to suggest that people seek such a basis only for what they value; they may also look for "biological" explanations for negative phenomena, perhaps to explain their sense of powerlessness before widespread or apparently intractable aspects of life.

2 What Does the Phenocopy Copy? Originals and Fakes in Biology

This chapter appears in slightly different form in *Psychological Reports 48,* 571–581 (1981). It is clear in retrospect that my assertion about the disappearance from academic writing of old-style nature-nurture-speak was mistaken. For another discussion of "potential," see Lewontin, Rose, & Kamin (1984). Lewontin's paper on ANOVA is here cited as (1974/1985) to avoid repetition of references, though the 1985 version is not the one I originally used.

Some discussion of Piaget's (1978) unorthodox take on phenocopies can be found in my more recent work (Oyama 1992, 1999). Waddington (1975) discusses theorists such as Richard Goldschmidt (1982), who saw evolution in developmental terms. One of the papers in Waddington's collection is called "The Evolution of Developmental Systems" and dates from the early 1940s. This was not the origin of my own use, but his emphasis on ontogeny was exemplary. It is too easy to forget that there is a long tradition of such thinking.

1 See Waddington (1975, p. 217) for comments on this kind of definition, and E. B. Ford (1965, pp. 28–29) on the error of attributing recessiveness and dominance to genes rather than to traits. Unless noted otherwise, I follow contemporary usage: *phenotype* to characterize the organism itself, *genotype* to refer to its genetic makeup.

2 Thiessen (1972, p. 113) observes that evolution is a conservative process, sometimes maintaining an advantageous phenotype through considerable turnover in genes.

3 This is the set of developmental outcomes of a genotype in different environments.

4 Correlated with degrees of relatedness—that is, with genetic differences.

5 Though Whelan does not seem to make these errors in interpreting heritability, this substitution does not seem likely to improve matters.

3 Ontogeny and the Central Dogma: Do We Need the Concept of Genetic Programming?

This chapter is based on a talk given at the 22nd Annual Minnesota Symposium on Child Psychology in 1987, and appeared in slightly different form in M. Gunnar & E. Thelen (Eds.) (1989) *Systems and development: Minnesota symposia on child psychology,* Vol. 22 (pp. 1–34). Hillsdale, NJ: Erlbaum. Further discussion of Tinbergen's (1963) four *whys* can be found in Klama (1988). In the light of more recent work (Griesemer, in press; Winther, in press), my mention in note 6 of Weismann's separation of development from transmission, historically an enormously important boundary, seems too stark. It is the reading of Weismann that has become dominant, however, and my point turns on just that dominant reading.

1 In the past I have not always distinguished clearly between the kind of interactionism I criticize as being inadequate and the kind I think overcomes those inadequacies. The former I often call "traditional" or "conventional" interactionism, the latter, real interactionism. But sometimes no modifier is used, and I have ungenerously expected the reader to know which meaning I intend. Others have avoided this problem by using another label (Lewontin, Rose & Kamin, 1984; Tobach & Greenberg, 1984). Although I shrink at coining yet another term, I offer *constructivist interactionism* as a name for my approach.

The notion of construction has its own problems, one of which is that it is often associated with environmental or social determinism, and so can imply arbitrary and unlimited variation. As should be clear from the Introduction and chapter 1, that is not what is intended here; what I wish to emphasize is the emergence of the phenotype, as well as its own active role in that emergence.

2 At least at the level of generality at which they are considered universal, so that there may be heritable variation in nose shape but not much in the *presence* of a nose.

3 The notion of programmatic control probably reflects the technocratic zeal of early systems thinkers (Haraway, 1981–1982; Taylor, 1988; Yoxen, 1981) better than it does the reciprocal, distributed control that characterizes a dynamic system. For descriptions of the latter, see M. C. Bateson (1972) and Fogel and Thelen (1987).

4 Instinct belongs to a whole set of concepts that mix evolutionary issues (the

behavior is the product of natural selection) with developmental ones (it arises spontaneously, with no guidance from the environment, and is resistant to variations in experience). One of my prime concerns is to show that an evolutionary perspective does not require this conflation.

5   Not too long ago I heard a developmental psychologist say that certain cognitive abilities of young children were "maturational." I asked her what she meant by that word. Evidently thinking I was objecting to an "extreme" position, she said, "Well, nothing's *completely* maturational." When I persisted in asking what the word *meant,* she finally retorted, with exemplary candor, "I mean it's present by the age of one, it's very complicated, and I don't want to think about it!"

6   Wilhelm Johannsen's genotype-phenotype distinction was intended to be an antidote to the deterministic, particulate notions of heredity as "transmission." Traits were not inherited, genes were. Johannsen opposed "the Weismannian mechanism and reductionism concealed within the corpuscular theory of heredity" (Sapp, 1987, pp. 37–50). Ironically, his distinction "offered geneticists the conceptual space or route by which they could bypass the organization of the cell, regulation by the internal and external environment of the organism, and the temporal and orderly sequences during development" (p. 49). With hindsight, I suggest that this misuse of Johannsen's idea was virtually inevitable in light of his limitation of heredity to only part of the causal system needed for development.

7   Modern evolutionists rely heavily on Weismann's separation of transmission from development. Brian Goodwin (1985) describes Weismann's "radically dualist" conception of the organism: a "generative and immortal germ plasm and a transient, mortal somatoplasm which was effectively the adult organism." Transmitted germ particles "stood in a specifically causal relation to a particular part of the organism" (p. 46). Goodwin also observes that with the advent of molecular biology this atomistic view was perpetuated with the concept of the genetic program.

8   It is our genecentrism that forces us to group phenomena of such different scales as "the environment." When DNA is the center of the universe, everything else is just envelope. Cohen (1979) speaks of the indifference of "DNA-is-God-and-RNA-is-his-prophet molecular biologists" and geneticists to the complexity of the egg; they tended to consider it merely a passive haploid (half a set of genes) awaiting penetration by another (one need hardly add *active*) haploid. Embryologists were more impressed by the role of egg structures. We see again the depth of Hamburger's nucleocytoplasmic gap. Cohen is interested in *reproduction,* and it is not an interest that can be satisfied by the equations of population genetics. It is just the rich complication of reproductive processes, Keller (1987) argues, that population

geneticists manage to ignore. In fact, the disembodying of genes facilitates the elision of some of the messier aspects of reproduction, among them the fact that organisms very rarely "reproduce themselves"—biparental contributions make "reproduction" a misnomer.

The integration of developmental processes renders problematic even the experimental separation of environment from genes. Raising genetically dissimilar organisms in the "same" environment and attributing their differences to the genes alone either ignores the effects of the different *effective* environment or manages to see them as proof of the organizing power of the genome (see note 9).

9 Wimsatt (1986) presents a developmental version of the distinction. I am in sympathy with much of his analysis, which overlaps my own. From the vantage point of the present project, however, his choice to retain *innateness* as a term for transgenerationally stable patterns is not ideal.

10 It is important to avoid the genetic imperialist impulse here. Scarr and McCartney (1983) point out the importance of experience in development, as well as the role of the child in influencing and selecting its experiences. Then they virtually eliminate the child by attributing agency to its genes: "*genotypes are the driving force behind development*"; the "impetus" for the experiences necessary for development comes "from the genotype" (p. 428). As I noted earlier, their insistence on the primacy of the gene is joined to talk of "developmental systems" that are radically different from the ones under discussion here.

4 Stasis, Development, and Heredity: Models of Stability and Change

This was based on remarks at a 1986 Liberty Fund Conference in Hawk's Cay, Florida. It appears under a shorter title and in slightly different form in M.-W. Ho & S. W. Fox (Eds.) (1988), *Evolutionary processes and metaphors* (pp. 255–274), London: Wiley. Neumann-Held (1999) takes the reassembly necessary to reproduce "genetic information," referred to in the section "Persistence and Reconstruction," and bases her redefinition of the gene on those processes.

1 The variational model is thus broader than the selectional one, including not only selection but phenomena like genetic drift as well (Sober 1985). In exploring the structure of these models Lewontin and Sober voice many of the criticisms presented here, and the former's concept of the interpenetration of organism and environment (Lewontin, 1982) fits well with my concept of the developmental system.

2 Sober (1984, pp. 153–155) substitutes "developmental" for "transforma-

tional." He considers the theories of Piaget and Chomsky to be "developmental" because of their emphasis on "preordained" sequences and "endogenous constraints." Not all developmental theories involve the unfolding of predestined form, of course, and although both Piaget and Chomsky are concerned with universals, their views of development are quite different.

It is the variational approach to evolution that is associated with stasis in these accounts, but the transformational model involves a kind of stasis as well. In psychology these preformationist connotations are present in instinct theories. They are also evident in traditional conceptions of maturation—the ontogenetic unfurling of preexisting forms (Oyama, 1982).

3 Sober argues that it is a mistake to speak of a metaphorical relationship between natural and artificial selection. Rather, he says, the latter is natural selection occurring in a particular niche (1984, p. 19). Even if one takes this position, however, it is reasonable to point out the special qualities of this niche—something Darwin himself was careful to do. If the image of a deliberating, manipulating breeder with fixed goals influences our thinking about evolution, it is perhaps worthwhile to reflect on this influence. Young (1985, chap. 4) argues that the anthropomorphism of Darwin's descriptions of natural selection had the paradoxical effect of increasing the acceptance of his ideas for just the wrong reasons: by making evolution seem to be the work of a creating agent.

4 Just as a ball seeks its natural position by rolling down an incline, so, by this logic, does a phenotype approach its natural condition by maturation. Waddington's (1957) image of an epigenetic landscape through which a ball-like organism rolls unfortunately lends itself all too easily to this model.

5 Research on the permeability of Weismann's barrier is both interesting and important (see Ho & Fox, 1988). While the reformulation I am proposing is entirely consistent with such permeability, it does not require or entail it. In fact, exclusive concentration on feedback to DNA structure risks legitimizing the tendency to define evolution in terms of genes alone.

6 This fact is easy to ignore when *evolved, biological,* and *physical* are treated as synonyms. Lest I imply a false dichotomy between body and mind, let me point out that both morphology and physiology often reflect variations in the activities and experiences of the organisms.

7 This statement probably doesn't do justice to much of animal behavior, but the point, I think, stands.

8 These questions are usually asked about genetic changes, but they are appropriate for nongenetic ones as well. Damping out is common in human affairs (traditions can weaken and disappear), and the change that fails to recur in offspring is a prototype of the "acquired" character.

Maintenance is seen in Douglas-Hamilton's (cited in Bonner, 1980, p. 177) account of radically altered behavioral patterns in elephants following intensive hunting. The formerly tame animals became highly aggressive and nocturnal and remained so four generations later. (Bonner 1980 gives other examples of transgenerational regularity; the abundant examples in his book, in fact, are perhaps more useful than his opposition of genetic and behavioral trait transmission.) Maintenance is also seen in Denenberg and Rosenberg's (1967) account of early handling of female rats; these experiences affected not just the females' behavior, but the behavior and weaning weight of their grandoffspring. In a discussion of the perpetuation of range selection, feeding preferences, and modes of predator avoidance in a wide range of vertebrates, Galef (1976) observes that the opposition of inheritance and individual acquisition of behavior had hindered recognition of such social processes.

Amplification is seen in Ho's (1984) description of cumulative cytoplasmic effects in fruit flies. Many social changes in humans show positive feedback characteristics; the rich may get richer and beget children who are richer still.

5 Ontogeny and Phylogeny: A Case of Meta-Recapitulation?

This chapter is based on a talk presented at "Philosophical Problems in Evolutionary Biology," a conference held in 1990 in Dunedin, New Zealand. It appears in slightly different form in P. E. Griffiths (Ed.) (1992), *Trees of life: Essays in philosophy of biology* (pp. 211–239), Dordrecht: Kluwer Academic Publishers. The habits of argumentation discussed here are very general ones. For some discussion of historical time scales, see anthropologist Tim Ingold's work on the "anatomically modern human" (1995, with commentaries and his reply, in Oyama et al. 1996). For analysis of the developmental constraints literature, see Amundson (1990, 1994). Godfrey-Smith (1996) presents a variety of views on internal and external causes.

1 Or, more globally, "the environment." To realize just how global most references to the environment are, consider the fact that "the environment" usually means "everything in the universe except the genes," and that in carving creation into these two segments one must conceptually excise the DNA from the cells in which it resides. See Oyama (1990a).

2 See Oyama (1982). Although the transformational model is formulated to explain change in a collection of entities, the focus in the present discussion is, of course, on change in an individual organism. One could think of organismic development in terms of transformational change at the level of

organs or tissues, but this would involve attributing the higher-level development to a heterogeneous assemblage of subsystems, and tissues exert a great deal of influence on each other. The organization among the subsystems would then present additional problems for a straightforward transformational account.

3 By this I mean only statistical typicality; no essential species nature is required for such probabilistic generalizations to be made.

4 See Skinner (1981). On neural selection, see Edelman and Mountcastle (1978); on the immune system, see Jerne (1967); on language, see Wexler and Culicover (1980); and for a sweeping view of selectionist explanation, as well as many references, see Piattelli-Palmarini (1989). It should be noted that there is a difference between selection from an actual array of objects or responses (organisms, neurons, operants) and selection as parameter setting, a difference that tends to be ignored when selectional models from neurobiology or immunology are mustered as support for the nativist project in the cognitive sciences.

5 This has the qualities of a good origin myth, but the story is more complex. For more appreciative views of Lamarck, see Ho and Saunders (1982) and Taylor (1987). For more historical detail, see Gruber (1981) and Jordanova (1984).

6 Or "ahistorical universals" (Kauffman, 1985, p. 171). Similarly, Ho and Saunders (1979, p. 590) declare that "a scientific study should consist in the delimitation of the necessities which underlie the process of evolution, *without recourse to contingencies.*" See also Goodwin (1982a).

7 The relation between artist and materials is being rethought, even in the popular press. Of an exhibit of Japanese sculpture a newsmagazine reports: "The exhibition rightly contends that its artists (or any artists, if you think about it) don't transform their materials so much as redirect them. They don't make everlasting objects out of inert and characterless stuff. . . . Instead, they highlight a few inherent qualities of their materials for a relatively brief moment in time" (Plagens, 1990, p. 64).

8 For classic critiques of nature-nurture dichotomizing, see Lehrman (1953, 1970). The distinction between genetic and environmental information is associated with Lorenz (1965). For discussions of this partitioning of information see Oyama (2000) and Johnston (1987).

9 Gray (1989, p. 803) maintains that constraints are considered primary. Although he and I read the literature differently, neither of us likes the insistence on designating one cause as dominant. The genes cannot, of course, define an array of possible outcomes independently of an array of developmental environments (each of which is actually an indefinitely long sequence of environments at various scales). As I mentioned earlier, further-

more, there is a certain vacuity to declaring that only those phenotypes that *can* develop *will* develop, and it makes no sense to place determinative power in only one set of interactants when every developmental outcome is jointly specified by genotype-environment (actually, organism-environment) pairings.

10 See Keller (1985) on the language of domination in science, and Sapp (1987) on the battles over the relative importance of the nucleus and the cytoplasm in development. The combatants in the nucleus-cytoplasm conflict deployed some of the same rhetorical strategies I describe here; there are insides and outsides even within the cell membrane. For more on intellectual politics, see chapter 11.

Although my comments here can be judged within the orthodox frameworks of developmental and evolutionary theories, they arise from a somewhat different one. It is possible to argue for an alternative framework without lapsing into a nihilistic relativism. To do so, however, involves some rather serious thinking about science. See Longino (1990) for some efforts in this direction.

11 Alberch (1982b) characterizes the developmental generation of a new bauplan as proceeding "autonomously from external environmental factors" (p. 23) and declares that "the evolution of developmental systems is characterised more by the internal structure of the developmental programme than by the external evolution of the environment" (p. 25).

12 For something like this move, see Webster and Goodwin (1982) and Goodwin (1982a); see also Alberch (1980). Alberch says that development is "crucial" in that "it defines the realm of the possible." Significantly, in describing macroevolution as an interaction between "production of morphological novelties (epigenetically determined) and differential extinction (environmentally determined)" (p. 664), he maintains the classic internal-external dichotomy, in which epigenesis is an internal process, independent of the environment, and selection, an external one, independent of development.

13 Bonner emphasizes the importance of other constituents of the germ cell, maintaining that a focus on nuclear DNA is too narrow. I agree, but find no warrant for stopping at the boundary of the cell.

14 Lewontin, Rose, and Kamin (1984, p. 275) also speak of "codevelopment of the organism and its environment." See Levins and Lewontin (1985) as well.

15 This sense of "developmental system" must therefore be distinguished from other uses of the phrase, such as Alberch's (1982a) decidedly internalist "developmental systems," which seem more or less equivalent to his "developmental programmes" (Alberch 1980, 1982b). The point should be

the *extension* of the concept of ontogeny, not the use of the traditional concept to curb the power of selection. Similarly, the definition of heredity presented here is significantly broader than the "hereditary apparatus" of Ho (1984). Consisting of the nucleus and cytoplasm, her "apparatus" is more restricted than the formulations in her later papers, as in her (1988a), in which she appears to argue for the inheritance of something like my developmental system.

Johnston and Turvey (1980, p. 152) capture the same interactive complex when they speak of the "co-implicative" relationship between organisms and their surrounds, as does Gray (1987b) with his "reciprocally constrained construction." Patten's (1982) "environs" may also refer to the same sort of inclusive complex. Whether these authors focus on development or on function (as in this last group of works), they have in common an interest in reducing the conceptual distance between organisms and their surroundings. See also Taylor (1987); Levins and Lewontin (1985); Lewontin, Rose, and Kamin (1984).

6 The Accidental Chordate: Contingency in Developmental Systems

This chapter is based on a talk given at the 1993 meeting of the International Society for the History, Philosophy, and Social Studies of Biology, at Brandeis University, in Waltham, Massachusetts. It appears in slightly different form in (1995) *South Atlantic Quarterly, 94*(2), 509–526. That paper is also reprinted in B. H. Smith and A. Plotnitsky (Eds.) (1997), *Mathematics, science, and postclassical theory* (pp. 118–133), Durham, NC: Duke University Press.

1 For discussion of the program metaphor, which is a commonplace in biology and in other fields, see Goodwin (1970); Nijhout (1990); and Oyama (2000).

Although the issue of progress or directionality in evolution is not closed, it tends to be discussed by evolutionists and philosophers in rather more rarified terms; see Nitecki (1989). The emphasis on (nonadaptive) chance in *Wonderful life* is part of a much larger argument about hierarchy in evolution and the adequacy of natural selection as an explanation for all levels of evolutionary change. For an outline of this argument, see Gould (1982).

2 Both ontogeny and phylogeny, of course, are historical phenomena, and it is precisely the conceptualization of history that is at issue here.

3 Compare this with the relations between object and context in Callon (1991), and between person and setting, in Lave (1986).

4 Notice the unspoken assumption of informational compressibility, of the possibility of giving a rule that is shorter than a "mere description" of what happens. On description and explanation, see Callon (1991) pp. 154–155.

5 These objections were raised at the conference at which an earlier version of this chapter was presented, and they are routinely raised in other contexts as well. There is more than a passing similarity between these objections and those answered, in somewhat different ways, in Latour (1991); B. H. Smith (1988, 1992); and Star (1991). The political and moral issues the authors address are sometimes intimately enmeshed with the very questions of biological essence and naturalness/normality/necessity that are at stake in the developmental biological and psychological literatures. See also chapter 10.

6 A case in point is the suggestion, made by Kim Sterelny, Kelly Smith and Michael Dickison (1996), that the developmental systems perspective leads to extending replicator status to non-genetic, even nonliving, constituents of a repeating life cycle. The terms derive from Dawkins's (1976) "replicators" and "vehicles," redefined and renamed "replicators" and "interactors" by David Hull (1988). Such terms, as Hull recognizes, tend to be associated with a particular style of explaining evolutionary change by the machinations of "selfish" entities maneuvering to ensure that replicas of themselves will appear in the next generation. See Griffiths and Gray (1994) for a developmental systems–style alternative.

7 Essentialism, Women, and War: Protesting Too Much, Protesting Too Little

This chapter is based on a talk given at the 1986 Genes & Gender conference in New York City. It appears in slightly different form in A. E. Hunter (Ed.) (1991), *Genes & gender VI, On peace, war, and gender* (pp. 64–76), New York: Feminist Press. It is also reprinted in M. M. Gergen & S. N. Davis (Eds.), (1997), *Toward a new psychology of gender: A reader* (pp. 521–532), London: Routledge Press; and in D. L. Hull & M. Ruse (Eds.) (1998), *The philosophy of biology* (pp 414–426), Oxford: Oxford University Press.

I refer in this chapter to a kind of feminist essentialism as "radical feminism." Alice Echols (1989) considers this instead to be part of the cultural feminism that she says supplanted the more politically oriented radical variety by the mid-1970s, although she says that *radical feminism* also continues to be used in the way I use it here. She notes too that an emphasis on the "eternal and unchanging" easily survives a switch from biological determinism to social constructionism (1989, p. 363). Meanwhile, the distinction between sex and gender remains significant in the literature on sexuality. This symmetry between social constructionism and many biological views is taken up especially in chapter 10.

For some more recent views on gender and war, see Barbara Ehrenreich's *Blood rites: Origins and history of the passions of war* (1997), and Ehrenreich and McIntosh (1997, June 9). Several papers in Mary Gergen and Sara Davis's (1997) are also relevant.

1   For a good discussion of the essentialist theme in feminism, see Jaggar (1983, chap. 5). See also Sayers (1982, p. 148). Bleier (1984, chap. 1) and Fausto-Sterling (1985, p. 195) also take up human sociobiology. Of these authors, Jaggar is perhaps most successful in transcending the biology-culture opposition, but all are aware of the mischief it has caused for scientists and nonscientists alike.

2   See S. Goldberg (1973) for an example, and Sayers (1982) for discussion.

3   I follow Jaggar (1983, chap. 5) and Sayers (1982) in using the term *radical* here. For the purposes of this chapter, it refers to a tendency to speak of essential feminine qualities in a positive, even celebratory way, rather than insisting on women's basic similarities to men.

4   See Money and Ehrhardt (1972) for a flawed but highly influential treatment of sex differences; see also the critiques by Bleier (1984) and Fausto-Sterling (1985); and see Klama (1988) on more general issues in aggression studies.

8 The Conceptualization of Nature: Nature as Design

This chapter is adapted from a talk given at a 1988 conference, "Biology as a Basis for Design," held at the Centro Luigi Bazzucchi in Perugia, Italy, and from a paper of the same name in W. I. Thompson (Ed.) (1991), *Gaia, vol. 2: Emergence: The science of becoming* (pp. 171–184), Hudson, NY: Lindisfarne Press.

Although the tone of this chapter is somewhat more personal and "site specific" than the others, it adverts to important issues that are treated in more detail elsewhere in the book. These include the political aspects of knowledge, explored in chapter 11, and strategies of causal analysis, discussed further in chapter 6. For a recent treatment of the subtleties of epistemology, feminist and otherwise, see B. H. Smith (in press).

1   Being of Japanese descent, I admit to a somewhat personal and exaggerated interest in this revisionist semantics.

2 See chapter 3. To the extent that this kind of mobile multiplicity fosters the appreciation of both particularity and commonality, it can, under the right circumstances, even serve to encourage empathy. Trying to see the consequences of one's actions through another's eyes, after all (perhaps even including the "eyes" of Nature), has long been an exercise for increasing moral and emotional responsiveness. As G. Bateson and M. C. Bateson write, "empathy is a discipline" (1987, p. 195). Such sensitivity can even give rise to the sense of extended responsibility so nicely embodied in the work on sustainable agriculture, energy-efficient design, and ecologically sound waste treatment (by people such as John and Nancy Todd, Wes Jackson, Sim Van der Ryn, and Amory and Hunter Lovins) that inspired this chapter.

3 On the relationship between the two, see Harding (1986). This is a complex and difficult problem; Harding believes it is possible, and necessary, to take a critical stance toward both.

4 On the importance in agriculture of nonscientists' knowledge, and on alternative research methods that use existing conditions in underdeveloped areas rather than requiring expensive and inaccessible equipment and techniques, see Levins (1981).

9 Bodies and Minds: Dualism in Evolutionary Theory

This chapter appears in slightly different form in L. R. Caporael and M. B. Brewer (Eds.) (1991), *Journal of Social Issues, 47*(3), 27–42. The current reproductive payoffs that were often prominently featured in sociobiology and that are central to the work described here are deemphasized by the evolutionary psychology of *The adapted mind* (Barkow, Cosmides, & Tooby, 1992), for instance, or Pinker's books (1994, 1997). This more recent body of work, however, is characterized by many of the difficulties discussed here (Griffiths, 1997; B. H. Smith, February 20, 1998; Sterelny, 1995). Sober and Wilson's most comprehensive treatment of altruism is their 1998 book.

1 The editors of a volume of papers by psychologists and biologists participating in the Bielefeld Interdisciplinary Project in the late 1970s wrote that *biological* usually means either "physiological" or "innate" (Immelmann, Barlow, Petrinovich, & Main, 1981). They also remarked on the temptation to equate "ultimate" factors with innateness and "proximate" ones with environmental control. (Each of these terms is itself used in various ways.) They adopted a third meaning for *biological:* "shaped by natural selection." Not surprisingly, there was considerable diversity in the way the term was

actually used by the various contributors, who sometimes resorted to traditional oppositions like biology versus experience or biology versus learning. I have reservations about the editors' definition, but it does have the virtue of cutting across most nature-nurture oppositions.

2 Such abuse of stepchildren would presumably be deserving of lesser punishment. Beckstrom also says, however, that the same evolutionary argument could justify *harsher* punishment for stepparents, on the grounds that they need stronger deterrents.

3 The assumption that evolution shows us what is desirable is not discussed here, but it has not vanished from the literature (see Masters, 1986). For other discussions of ambiguity in the nature-nurture complex, and other examples of questionable inference, see Johnston (1987); Kitcher (1985). See also Wimsatt (1986) on innateness.

4 These are not the same thing, incidentally. *Inevitability* refers to certainty of occurrence; *immutability* means nonsusceptibility to influence once present. Both are different from *universality.* See Lehrman (1970) for some classic remarks on heritability and fixity.

5 This is the idea that individuals act for the evolutionary "good" of the population or species. See D. S. Wilson (1989) for a rebuttal of this attitude of blanket rejection and a critique of "cheap individualism," as well as a defense of the legitimacy of exploring supraindividual levels of selection. See also Sober (1988) for discussion of levels of selection and the difference between "vernacular," or psychological, altruism and evolutionary altruism.

10 How Shall I Name Thee? The Construction of Natural Selves

This chapter appears in slightly different form in B. Bradley and W. Kessen (Eds.) (1993), *Theory and Psychology, 3,* 471–496. John Shotter's more recent work (1993) is relevant to my discussion here of the informational gene and compucentric cognitivism's symbol-manipulating subject. My doubts about "cognitivism" (a term used in Varela, Thompson, & Rosch, 1991; see also Hendriks-Jansen, 1996; and van Gelder & Port, 1995) do not extend to cognitive psychology in general.

1 See N. Rose (1985); and Henriques et al. (1984). We may be formed as much by what we reject as by what we embrace: Struggling to confound the expectations of an important "other" is nevertheless to be affected.

2 Like the physiological processes that maintain life, moment-by-moment social interactions, in which our positions and senses of ourselves may shift rapidly, may seem too evanescent to be considered developmental. I do not think a clear line can be drawn between these daily processes and develop-

mental ones as they are usually understood, however; at some point change will seem marked enough that an observer will say that the person is now different.

3 Although this view is typically dated from the seventeenth century, I do not mean to imply total discontinuity from earlier periods. The separation between reason and passion, for instance, or between form and matter, two other troublesome polarities, predates Descartes. Shweder (1990, p. 19) says that the Platonic assumptions of cognitivism indicate a desire "for something abstract and bleached and really real."

4 There is a startling disanalogy: While the subject in cognitive science is the subject of knowledge (or, better, "information," because knowledge implies content and meaning), not action, the genetic "subject" both knows and acts, and so seems the more nearly complete subject. Yet, computer-talk dominates in both cases, and abstract representations are more prominent than material interactions. Once properly contextualized in its cellular milieu, however, the gene interacts by virtue of its three-dimensional structure, without benefit of representations.

5 See chapter 4 for comments on Williams's dematerialization of the gene into a "weightless package" of information.

6 I suspect we are more apt to use causal explanations when we do not understand, or cannot accept, another's reasons. This may help to explain why reasons and causes have so often been treated as mutually exclusive.

7 My use of "person" to indicate both subjectivity (awareness, including self-awareness) and accountable agency (capacity for intentional action in a moral order) is somewhat similar to Dennett's (1981) personal stance, which combines the intentional and moral attitudes, and to Shotter's (1984, p. 185) social accountability stance. The latter's distinction between acting as one requires and as one's circumstances require (pp. 213–214) is perhaps not as useful. I am thus less skeptical of *some notion* of agency than are Henriques et al. (1984), for instance; in fact, I intend by this term to include just the sorts of attempts, assertions, and situated acts they discuss.

8 That this is necessary is evident from the way Scarr and McCartney (1983, pp. 428, 433) first speak of active children, then go on to attribute *real* agency to the genes that control development. Genes are even said to find some experiences more "compatible" than others (see chapter 3).

9 The choice captures England's movement from a moral view of homosexuality to a more medical one. Hodges interprets Turing's choice of estrogen (to suppress libido) as one in favor of mind (the possibility of continuing his intellectual work) over body (1983, p. 473), but the apparent suicide suggests they were indivisible.

10 I hope it is clear that my broadening of the concepts of development and biology, far from being a bid for disciplinary dominance, defuses the terms, by diffusing them (chapter 5). That the construction of persons is part of more general developmental processes does not make it the special purview of evolutionary biology, genetics, or any other traditional division of the field. Formative power is not concentrated in a single factor, and nature may not be invoked to decide what is morally permissible, inevitable or beyond the reach of action.

11 Evolutionary and Developmental Formation: Politics of the Boundary

This chapter is based on two talks given in 1996, one at "Developmental Systems, Competition and Cooperation in Sociobiology and Economics," a conference sponsored by the Stiftung Forschungsinstitut für Philosophie Hannover, in Marienrode, Germany; and the other at "Evolution and Human Behavior," a conference at Stanford University. It appears in slightly different form in P. Koslowski (Ed.), (1999), *Sociobiology and bioeconomics: The theory of evolution in biological and economic theory* (pp. 79–104), Berlin: Springer-Verlag.

1 Neutralism refers to evolution by random changes in the frequencies of selectively neutral genes, in contrast with the natural selection of genetic variants "for" their contribution to survival and reproduction (Kimura, 1992). Godfrey-Smith (1996) supplies more examples of creative line drawing.

2 Sociobiologists have sometimes complained of being unjustly characterized as apologists for retrograde politics, for instance, as we saw in chapter 9. Not that one can avoid being called a Nazi or anything else by taking a critical stance toward these ideas; culture theorists and relativists of every stripe hear the charge, too. As I suggested above, those who seek to broaden evolutionary theory now risk being labeled creationists. The point here is not that one can totally control the names one is called, but that it is sometimes possible to gain some insight into the ways name-calling works.

3 The use of *game* here obviously does not imply lack of seriousness.

4 Dennett's (1995) autonomous, hyperactive genes and memes, however, seem rhetorical aids to his denial that humans have intrinsic or original intentionality.

5 Sterelny, Smith, and Dickison (1996) do not balk here. Taking the idea of a replicator more seriously, or at least consistently, than its author does, they apply it to extraorganismic entities.

6 See also Depew and Weber (1996); and see Weber and Depew (1999) for an attempt to rework selection.

7 These are not mutually exclusive classes, especially when one thinks in terms of systems in which consequences are causally important. Think of the idea of feedback. This does not change the point about investigative focus.

References

Alberch, P. (1980). Ontogenesis and morphological diversification. *American Zoologist, 22,* 653–667.

Alberch, P. (1982a). Developmental constraints in evolutionary processes. In J. T. Bonner (Ed.), *Evolution and development* (pp. 313–332). Berlin: Springer-Verlag.

Alberch, P. (1982b). The generative and regulatory roles of development in evolution. In D. Mossakowski & G. Roth (Eds.), *Environmental adaptation and evolution* (pp. 19–34). Stuttgart: Gustav Fischer.

Alberts, B., Bray, D., Lewis, J., Raff, M., Roberts, K., & Watson, J. D. (1983). *Molecular biology of the cell.* New York: Garland.

Alexander, R. D. (1979). *Darwinism and human affairs.* Seattle: University of Washington Press.

Alexander, R. D. (1987). *The biology of moral systems.* New York: Aldine de Gruyter.

Alland, A., Jr. (1973). *Evolution and human behavior.* (2nd ed.). New York: Anchor Books.

Amundson, R. (1990). Doctor Dennett and Doctor Pangloss: Perfection and selection in psychology and biology. *Behavioral and Brain Sciences, 13,* 577–584.

Amundson, R. (1994). Two concepts of constraint: Adaptationism and the challenge from developmental biology. *Philosophy of Science, 61,* 556–578.

Anastasi, A. (1958). Heredity, environment, and the question "how"? *Psychological Review, 65,* 197–208.

Andresen, J. T. (1992). The contemporary linguist meets the postmodernist. *Beiträge zur Geschichte der Sprachwissenschaft, 2,* 213–223.

Angier, N. (1993, June 29). Researchers track pivotal pathway that makes cells divide. *New York Times,* pp. C1, C10.

Anonymous. (1985, October 15). Results of rat tests stir concern for astronauts. *New York Times,* p. C8.

Barash, D. (1979). *The whisperings within.* New York: Harper and Row.

Barash, D., & Lipton, J. E. (1982). *Stop nuclear war!* New York: Grove Press.

Barkow, J., Cosmides, L., & Tooby, J. (Eds). (1992). *The adapted mind.* Oxford: Oxford University Press.

Bateson, G. (1972). *Steps to an ecology of mind.* New York: Ballantine Books.

Bateson, G. (1979). *Mind and nature: A necessary unity.* Toronto: Bantam Books.

Bateson, G., & Bateson, M. C. (1987). *Angels fear: Towards an epistemology of the sacred.* New York: Macmillan.

Bateson, M. C. (1972). *Our own metaphor.* New York: Knopf.

Bateson, M. C. (1987, November 7). Talk presented at the Fifteenth Anniversary Meeting of the Lindisfarne Association, New York City.

Bateson, P. [P. G.] (1978). [Review of the book *The selfish gene*]. *Animal Behaviour, 26,* 316–318.

Bateson, P. P. G. (1979). How do sensitive periods arise and what are they for? *Animal Behaviour, 27,* 470–486.

Bateson, P. [P. G.] (1983). Genes, environment and the development of behaviour. In T. R. Halliday & P. J. B. Slater (Eds.), *Animal behaviour: Vol. 3. Genes, development, and learning* (pp. 52–81). Oxford: Blackwell.

Bateson, P. P. G. (1984). Genes, evolution, and learning. In P. Marler & H. S. Terrace (Eds.), *The biology of learning* (pp. 75–88). Berlin: Springer-Verlag.

Bateson, P. P. G. (1985). Problems and possibilities in fusing developmental and evolutionary thought. In G. Butterworth, J. Rutowska & M. Scaife (Eds.), *Evolution and developmental psychology* (pp. 3–21). Sussex: Harvester Press.

Bateson, P. P. G. (1986). Sociobiology and human politics. In S. Rose & L. Appignanesi (Eds.), *Science and beyond* (pp. 79–99). Oxford: Blackwell.

Bateson, P. P. G. (1987). Biological approaches to the study of behavioural development. *International Journal of Behavioral Development, 10,* 1–22.

Bateson, P. [P. G.] (1988a). The active role of behaviour in evolution. In M.-W. Ho & S. W. Fox (Eds.), *Evolutionary processes and metaphors* (pp. 191–207). London: Wiley.

Bateson, P. [P. G.] (1988b). The biological evolution of cooperation and trust. In D. Gambetta (Ed.), *Trust: Making and breaking cooperative relations* (pp. 14–30). Oxford: Blackwell.

Bateson, P. [P. G.] (1991). Are there principles of behavioural development? In P. Bateson (Ed.), *The development and integration of behaviour* (pp. 19–39). Cambridge: Cambridge University Press.

Bateson, P. [P. G.], and Martin, P. (1999). *Design for a life.* London: Cape.

Beach, F. A. (1955). The descent of instinct. *Psychological Review. 62,* 401–410.

Beckstrom, J. H. (1985). *Sociobiology and the law.* Urbana: University of Illinois Press.

Bellah, R. N., Masden, R., Sullivan, W. M., Swidler, A., & Tipton, S. M. (1985). *Habits of the heart.* Berkeley: University of California Press.

Bem, D. J. (1996). Exotic becomes erotic: A developmental theory of sexual orientation. *Psychological Review, 103*, 320–335.

Berlin, F. S. (1988). Issues in the exploration of biological factors contributing to the etiology of the "sex offender," plus some ethical considerations. In R. A. Prentky & V. L. Quinsey (Eds.), *Human sexual aggression. Annals of the New York Academy of Sciences, 528*, 183–192. New York: New York Academy of Sciences.

Bleier, R. (1984). *Science and gender: A critique of biology and its theories on women*. New York: Pergamon Press.

Block, N. J., & Dworkin, G. (Eds.). (1976). *The IQ controversy*. New York: Pantheon.

Bonner, J. T. (1974). *On development*. Cambridge: Harvard University Press.

Bonner, J. T. (1980). *The evolution of culture in animals*. Princeton: Princeton University Press.

Bordo, S. (1987). The Cartesian masculinization of thought. In S. Harding & J. F. O'Barr (Eds.) (1987), *Sex and scientific inquiry* (pp. 247–264). Chicago: University of Chicago Press. (Reprinted from *Signs, 11*(3), 1986.)

Boyd, R., & Richerson, P. J. (1985). *Culture and the evolutionary process*. Chicago: University of Chicago Press.

Brandon, R. N. (1990). *Adaptation and environment*. Princeton: Princeton University Press.

Brandon, R. N., & Antonovics, J. (1995). The coevolution of organism and environment. In G. Wolters and J. G. Lennow (in collaboration with P. McLaughlin) (Eds.). *Concepts, theories, and rationality in the biological sciences* (pp. 211–232). Pittsburgh: University of Pittsburgh Press.

Buss, D. M. (1984). Evolutionary biology and personality psychology: Towards a conception of human value and individual differences. *American Psychologist, 39*, 1135–1147.

Cahn, S. M. (1967). Chance. In P. Edwards (Ed.), *Encyclopedia of philosophy* (Vol. 2) (pp. 73–75). New York: Macmillan and the Free Press.

Cairns, R. B. (1979). *Social development: The origins and plasticity of interchanges*. San Francisco: W. H. Freeman.

Callon, M. (1986). Some elements of a sociology of translation: Domestication of the scallops and the fishermen of St Brieuc Bay. In J. Law (Ed.), *Power, action, and belief: A new sociology of knowledge? Sociological Review Monograph 32* (pp. 196–233). London: Routledge & Kegan Paul.

Callon, M. (1991). Techno-economic networks and irreversibility. In J. Law (Ed.), *A sociology of monsters: Essays on power, technology, and domination, Sociological Review Monograph 38* (pp. 132–161). London: Routledge & Kegan Paul.

Caporael, L. R. (1997). The evolution of truly social cognition: The core configuration model. *Personality and Social Psychology Review, 1,* 276–298.

Caporael, L. R., Dawes, R. M., Orbell, J. M., & van de Kragt, A. J. C. (1989). Selfishness examined: Cooperation in the absence of egoistic incentives. *Behavioral and Brain Sciences, 12,* 683–739.

Caro, T. M., & Bateson, P. [P. G.] (1986). Organization and ontogeny of alternative tactics. *Animal Behaviour, 34,* 1483–1499.

Churchill, F. B. (1974). William Johannsen and the genotype concept. *Journal of the History of Biology, 7*(1), 5–30.

Clarke, A. D. B. (1978). Predicting human development: Problems, evidence, implications. *Bulletin of the British Psychological Society, 31,* 249–258.

Cohen, J. (1979). Maternal constraints on development. In D. R. Newth & M. Balls (Eds.), *Maternal effects in development* (pp. 1–28). Cambridge: Cambridge University Press.

Cohen, J. (1989). *The privileged ape.* Park Ridge, NJ: Parthenon.

Connor, J. M., Schackman, M., & Serbin, L. A. (1978). Sex-related differences in response to practice on a visual-spatial test and generalization to a related test. *Child Development, 49,* 24–29.

Crawford, C. B. (1987). Sociobiology: Of what value to psychology? In C. Crawford, M. Smith, & D. Krebs (Eds.), *Sociobiology and psychology* (pp. 3–30). Hillsdale, NJ: Erlbaum.

Crawford, C. B., Smith, M. S., & Krebs, D. (Eds.). (1987). *Sociobiology and psychology.* Hillsdale, NJ: Erlbaum.

Crick, F. (1957). On protein synthesis. *Symposium of the Society of Experimental Biology, 12,* 138–163.

Cushman, P. (1991). Ideology obscured. *American Psychologist, 46,* 206–219.

Daly, M., & Wilson, M. (1988). *Homicide.* New York: Aldine de Gruyter.

Dawkins, R. (1976). *The selfish gene.* Oxford: Oxford University Press.

Dawkins, R. (1982). *The extended phenotype: The gene as the unit of selection.* Oxford: Oxford University Press.

Dawkins, R. (1986). *The blind watchmaker.* New York: Norton.

de Catanzaro, D. (1987). Evolutionary pressures and limitations to self-preservation. In C. Crawford, M. Smith, & D. Krebs (Eds.), *Sociobiology and psychology* (pp. 311–333). Hillsdale, NJ: Erlbaum.

Denenberg, V. H., & Rosenberg, K. M. (1967). Nongenetic transmission of information. *Nature, 216,* 549–550.

Dennett, D. C. (1981). Mechanism and responsibility. In D. C. Dennett, *Brainstorms: Philosophical essays on mind and psychology* (pp. 233–255). Cambridge: MIT Press/Bradford.

Dennett, D. C. (1984). *Elbow room: Varieties of free will worth wanting.* Cambridge: MIT Press/Bradford.

Dennett, D. C. (1995). *Darwin's dangerous idea.* New York: Simon & Schuster.

Depew, D. J., & Weber, B. H. (1996). *Darwinism evolving: Systems dynamics and the genealogy of natural selection.* Cambridge: MIT Press/Bradford.

Dillon, L. I. (1983). *The inconstant gene.* New York: Plenum Press.

Doyle, R. M. (1991/1997). *On beyond living: Rhetorical transformations of the life sciences.* Stanford: Stanford University Press.

Dray, W. H. (1967). Determinism in history In P. Edwards (Ed.), *Encyclopedia of philosophy* (Vol. 2) (pp. 373–378). New York: Macmillan and Free Press.

Durham, W. H. (1979). Toward a coevolutionary theory of human biology and culture. In N. A. Chagnon & W. Irons (Eds.), *Evolutionary biology and human social behavior: An anthropological perspective* (pp. 4–39). North Scituate, MA: Duxbury Press.

Dwyer, P. D. (1984). Functionalism and structuralism: Two programmes for evolutionary biologists. *American Naturalist, 124,* 745–750.

Dyke, C., & Depew, D. (1988). Should natural selection be an explanation of last resort? Well, maybe not the last resort, but . . . *Rivista de Biologia/Biology Forum, 81,* 115–129.

Echols, A. (1989). *Daring to be bad: Radical feminism in America 1967–1975.* Minneapolis: University of Minnesota Press.

Edelman, G. M., & Mountcastle, V. B. (1978). *The mindful brain.* Cambridge: MIT Press.

Ehrenreich, B. (1997). *Blood rites: Origins and history of the passions of war.* New York: Holt.

Ehrenreich, B., & McIntosh, J. (1997, June 9). The new creationism: Biology under attack. *Nation,* pp. 111, 116.

Elster, J. (1989). *Nuts and bolts for the social sciences.* Cambridge: Cambridge University Press.

Erlenmeyer-Kimling, L. (1975). Commentary 1: Nature-nurture redefined. In K. W. Schaie, V. E. Anderson, G. E. McClearn & J. Money (Eds.), *Developmental human behavior genetics* (pp. 25–31). Lexington, MA: Heath.

Fausto-Sterling, A. (1985). *Myths of gender: Biological theories about women and men.* New York: Basic Books.

Fausto-Sterling, A. (in press). *Sexing the body: Gender politics and the construction of human sexuality.* New York: Basic Books.

Fishbein, H. D. (1976). *Evolution, development, and children's learning.* Santa Monica, CA: Goodyear.

Fogel, A. (1993). *Developing through relationships.* Chicago: University of Chicago Press.

Fogel, A., & Thelen, E. (1987). Development of early expressive and communicative action: Reinterpreting the evidence from a dynamic systems perspective. *Developmental Psychology, 23,* 747–761.

Fontana, W., & Buss, L. W. (1996). The barrier of objects: From dynamical systems to bounded organizations. In J. Casti & A. Karlqvist (Eds.), *Boundaries and barriers* (pp. 55–115). Reading, MA: Addison-Wesley.

Ford, D. H., & Lerner, R. M. (1992). *Developmental systems theory: An integrative approach.* Newbury Park, CA: Sage.

Ford, E. B. (1965). *Genetic polymorphism.* Cambridge: MIT Press.

Freedman, D. G. (1979). *Human sociobiology: A holistic approach.* New York: Free Press.

Frodi, A. M., & Lamb, M. E. (1978). Sex differences in responsiveness to infants: A developmental study of psychophysiological and behavioral responses. *Child Development, 49,* 1182–1188.

Fujimura, J. H. (1992). Crafting science: Standardized packages, boundary objects, and "translation." In A. Pickering (Ed.), *Science as practice and culture* (pp. 168–211). Chicago: University of Chicago Press.

Fuller, J. L. (1987). What can genes do? In C. Crawford, M. Smith, & D. Krebs (Eds.), *Sociobiology and psychology* (pp. 147–174). Hillsdale, NJ: Erlbaum.

Furth, H. G. (1974). Two aspects of experience in ontogeny: Development and learning. *Advances in Child Development and Behavior, 9,* 47–67.

Galef, B. G., Jr. (1976). Social transmission of acquired behavior: A discussion of tradition and social learning in vertebrates. In J. S. Rosenblatt, R. A. Hinde, E. Shaw, & C. Beer (Eds.), *Advances in the study of behavior, 6* (pp. 77–100). New York: Academic Press.

Galef, B. G., Jr. (1988). Imitation in animals: History, definition, and interpretation of data from the psychological laboratory. In T. Zentall and B. G. Galef, Jr. (Eds.), *Comparative social learning* (pp. 3–28). Hillsdale, NJ: Erlbaum.

Gandelman, R., vom Saal, F. S., & Reinisch, J. M. (1977). Contiguity to male fetuses affects morphology and behavior in female mice. *Nature, 266,* 722–723.

Gelman, D., Foote, D., Barrett, T., & Talbot, M. (1992, February 24). Born or bred? *Newsweek,* pp. 46–53.

Gergen, K. J. (1985). The social constructionist movement in modern psychology. *American Psychologist, 40,* 266–275.

Gergen, K. J. (1990). Social understanding and the inscription of self. In J. W. Stigler, R. A. Shweder & G. Herdt (Eds.), *Cultural psychology: Essays on comparative human development* (pp. 569–606). Cambridge: Cambridge University Press.

Gergen, M. M., & Davis, S. N. (Eds.). (1997). *Toward a new psychology of gender: A reader.* London: Routledge.

Giddens, A. (1987). *Social theory and modern sociology.* Stanford: Stanford University Press.

Gilbert, S. F., Opitz, J. M., & Raff, R. A. (1996). Resynthesizing evolutionary and developmental biology. *Developmental Biology, 173,* 357–372.

Godfrey-Smith, P. (1996). *Complexity and the function of mind in nature.* Cambridge: Cambridge University Press.

Godfrey-Smith, P. (1999). Genes and codes: Lessons from the philosophy of mind? In V. Hardcastle (Ed.), *Biology meets psychology: Constraints, conjectures, connections* (pp. 305–331). Cambridge: MIT Press.

Goffman, E. (1959). *The presentation of self in everyday life.* New York: Doubleday.

Goldberg, R. (1943a). Get one of our patent fans and keep cool. In T. Craven (Ed.), *Cartoon cavalcade* (p. 214). New York: Simon and Schuster.

Goldberg, R. (1943b). Inventions of Professor Lucifer Butts. In T. Craven (Ed.), *Cartoon cavalcade* (p. 235). New York: Simon and Schuster. (Reprinted from *Colliers Magazine,* 1932.)

Goldberg, S. (1973). *The inevitability of patriarchy.* New York: Morrow.

Goldschmidt, R. (1940/1982). *The material basis of evolution.* New Haven: Yale University Press. (Original work published in 1940.)

Goodwin, B. C. (1970). Biological stability. In C. H. Waddington (Ed.), *Towards a theoretical biology* (Vol. 3) (pp. 1–17). Chicago: Aldine.

Goodwin, B. C. (1982a). Biology without Darwinian spectacles. *Biologist, 29,* 108–112.

Goodwin, B. C. (1982b). Genetic epistemology and constructionist biology. *Revue Internationale de Philosophie, 142–143,* 527–548.

Goodwin, B. C. (1984). A relational or field theory of reproduction and its evolutionary implications. In M.-W. Ho & P. T. Saunders (Eds.), *Beyond neo-Darwinism* (pp. 219–241). London: Academic Press.

Goodwin, B. C. (1985). Constructional biology. In G. Butterworth, J. Rutowska, & M. Scaife (Eds.), *Evolution and developmental psychology* (pp. 45–66). Sussex: Harvester Press.

Goodwin, B. C. (1989). Evolution and the generative order. In B. Goodwin & P. Saunders (Eds.), *Theoretical biology* (pp. 89–100). Edinburgh: Edinburgh University Press.

Gordon, D. (in press). The evolution of social behavior. In R. Singh, C. Krimbas, D. Paul, & J. Beatty (Eds.) *Thinking about evolution: Historical, philosophical, and political perspectives.* Cambridge: Cambridge University Press.

Gottlieb, G. (1970). Conceptions of prenatal behavior. In L. R. Aronson, E. Tobach, D. S. Lehrman, & J. S. Rosenblatt (Eds.), *Development and evolution of behavior: Essays in memory of T. C. Schneirla* (pp. 111–137). San Francisco: W. H. Freeman.

Gottlieb, G. (1971). *Development of species identification in birds.* Chicago: University of Chicago Press.

Gottlieb, G. (1976). Conceptions of prenatal development: Behavioral embryology. *Psychological Review, 83,* 215–234.

Gottlieb, G. (1978). Development of species identification in ducklings: IV. Change in species-specific perception caused by auditory deprivation. *Journal of Comparative and Physiological Psychology, 92,* 375–387.

Gottlieb, G. (1992). *Individual development and evolution.* Oxford: Oxford University Press.

Gottlieb, G. (1997). *Synthesizing nature-nurture: Prenatal roots of instinctive behavior.* Mahwah, NJ: Erlbaum.

Gould, J. L. (1982). *Ethology: The mechanisms and evolution of behavior.* New York: Norton.

Gould, J. L. (1985). Reply to letters from readers. *Sciences, 25*(6), 15.

Gould, S. J. (1977). *Ontogeny and phylogeny.* Cambridge: Harvard University Press/Belknap.

Gould, S. J. (1982). Is a new and general theory of evolution emerging? In J. M. Smith (Ed.), *Evolution now* (pp. 129–145). San Francisco: W. H. Freeman.

Gould, S. J. (1989a). *Wonderful life: The Burgess shale and the nature of history.* New York: Norton.

Gould, S. J. (1989b). A developmental constraint in Cerion, with comments on the definition and interpretation of constraint in evolution. *Evolution, 43,* 516–539.

Gould, S. J. (1992). Ontogeny and phylogeny—revisited and united. *BioEssays, 14,* 275–279.

Gould, S. J., & Lewontin, R. C. (1979). The spandrels of San Marco and the Panglossian paradigm: A critique of the adaptationist programme. *Proceedings of the Royal Society of London B 205,* 581–598.

Gray, R. D. (1987a). Beyond labels and binary oppositions: What can be learnt from the nature/nurture dispute? *Rivista di Biologia/Biology Forum, 80,* 192–196.

Gray, R. D. (1987b). Faith and foraging: A critique of the paradigm argument from design. In A. C. Kamil, J. R. Krebs, & H. R. Pulliam (Eds.), *Foraging behaviour* (pp. 69–140). New York: Plenum Press.

Gray, R. D. (1988). Metaphors and methods: Behavioural ecology, panbiogeography and the evolving synthesis. In M.-W. Ho & S. W. Fox (Eds.), *Evolutionary processes and metaphors* (pp. 209–242). London: Wiley.

Gray, R. D. (1989). Oppositions in panbiogeography: Can the conflicts between selection, constraint, ecology, and history be resolved? *New Zealand Journal of Zoology, 16,* 787–806.

Gray, R. D. (1992). Death of the gene: Developmental systems strike back. In P. Griffiths (Ed.). *Trees of life: Essays in philosophy of biology* (pp. 165–209). Boston: Kluwer.

Grehan, J. R., & Ainsworth, R. (1985). Orthogenesis and evolution. *Systematic Zoology, 34,* 174–192.

Griesemer, J. (in press). The informational gene and the substantial body: On the generalization of evolutionary theory by abstraction. In N. Cartwright & M. Jones (Eds.), *Varieties of idealisation,* Poznan Studies in the Philosophy of the Sciences and the Humanities (Leszek Nowak, series ed.). Amsterdam: Rodopi.

Griesemer, J., & Wimsatt, W. (1989). Picturing Weismannism: A case study of conceptual evolution. In M. Ruse (ed.), *What the philosophy of biology is: Essays for David Hull* (pp. 75–137). Dordrecht: Kluwer.

Griffiths, P. E. (1997). *What emotions really are: The problem of psychological categories.* Chicago: University of Chicago Press.

Griffiths, P. E., & Gray, R. D. (1994). Developmental systems and evolutionary explanation. *Journal of Philosophy, 91,* 277–304.

Griffiths, P. E., & Gray, R. D. (1997). Replicator II: Judgement day. *Biology and Philosophy, 12,* 471–492.

Griffiths, P. E., & Knight, R. D. (1998). What is the developmentalist challenge? *Philosophy of Science, 65,* 253–258.

Griffiths, P. E., & Neumann-Held, E. M. (1999). The many faces of the gene. *BioScience, 49,* 656–662.

Gruber, H. E. (1981). *Darwin on man: A psychological study of scientific creativity* (2nd ed.). Chicago: University of Chicago Press.

Gruber, H. E., & Sehl, I. A. (1984). Transcending relativism and subjectivism: Going beyond the information I am given. In W. Callebaut, S. E. Cozzens, B. P. Lécuyer, A. Rip, & J. P. Van Bendegem (Eds.), *George Sarton Centennial* (pp. 57–60). Ghent: Communication & Cognition.

Gruter, M. (1986). Ostracism on trial: The limits of individual rights. *Ethology and Sociobiology, 7,* 271–279.

Gruter, M., & Masters, R. D. (1986). Ostracism as a social and biological phenomenon: An introduction. *Ethology and Sociobiology, 7,* 149–158.

Hailman, J. [P.] (1969). How an instinct is learned. *Scientific American, 221*(6), 98–106.

Hailman, J. P. (1982). Evolution and behavior: An iconoclastic view. In H. C. Plotkin (Ed.), *Learning, development, and culture* (pp. 205–254). Chichester, NY: Wiley.

Hamburger, V. (1980). Embryology and the modern synthesis in evolutionary theory. In E. Mayr & W. B. Provine (Eds.), *The evolutionary synthesis* (pp. 97–112). Cambridge: Harvard University Press.

Haraway, D. J. (1981–1982). The high cost of information in post–World War II evolutionary biology: Ergonomics, semiotics, and the sociobiology of communication systems. *Philosophical Forum, 13*(2–3), 244–278.

Haraway, D. [J.] (1985a). In the beginning was the word: The genesis of biological theory. *Signs 6*, 469–481.

Haraway, D. [J.] (1985b). A manifesto for cyborgs: Science, technology, and socialist feminism in the 1980s. *Socialist Review, 80*, 65–107.

Haraway, D. [J.] (1987). Animal sociology and a natural economy of the body politic, Part 1: A political physiology of dominance. In S. Harding & J. F. O'Barr (Eds.), *Sex and scientific inquiry* (pp. 217–232). Chicago: University of Chicago Press. (Reprinted from *Signs, 4*, 1978.)

Haraway, D. J. (1989). *Primate visions: Gender, race, and nature in the world of modern science*. New York: Routledge.

Haraway, D. J. (1991). *Simians, cyborgs, and women: The reinvention of nature*. New York: Routledge.

Harding, S. (1986). *The science question in feminism*. Ithaca: Cornell University Press.

Harding, S., & O'Barr, J. F. (Eds.). (1987). *Sex and scientific inquiry*. Chicago: University of Chicago Press.

Harré, R. (1991). The discursive production of selves. *Theory and Psychology, 1*, 51–63.

Harris, M. (1983, July/August). Margaret and the giant-killer. *Sciences, 23* (4), pp. 18–21.

Hendriks-Jansen, H. (1996). *Catching ourselves in the act*. Cambridge: MIT Press/Bradford.

Henriques, J., Hollway, W., Urwin, C., Venn, C., & Walkerdine, V. (1984). *Changing the subject*. London: Methuen.

Herrnstein, R. J. (1989). Darwinism and behaviourism: Parallels and intersections. In A. Grafen (Ed.), *Evolution and its influence* (pp. 35–61). London: Clarendon.

Hinde, R. A. (1968). Dichotomies in the study of development. In J. M. Thoday & A. S. Parkes (Eds.), *Genetic and environmental influences on behaviour* (pp. 3–14). New York: Plenum.

Hinde, R. A. (1987). *Individuals, relationships, and culture*. Cambridge: Cambridge University Press.

Hinde, R. A., & Bateson, [P. G.]. (1984). Discontinuities versus continuities in behavioural development and the neglect of process. *International Journal of Behavioral Development, 7*, 129–143.

Hinde, R. A., & Stevenson-Hinde, J. (1973). *Constraints on learning: Limitations and predispositions*. New York: Academic Press.

Hirsch, J. (1970/1976). Behavior-genetic analysis and its biosocial consequences. In N. J. Block & G. Dworkin (Eds.). *The IQ controversy* (pp. 156–178). New York: Pantheon. (Reprinted from *Seminars in Psychiatry, 2*, 1970.)

Ho, M.-W. (1984). Environment and heredity in development and evolution. In

M.-W. Ho & P. T. Saunders (Eds.), *Beyond neo-Darwinism: Introduction to the new evolutionary paradigm* (pp. 267–289). London: Academic Press.

Ho, M.-W. (1988a). On not holding nature still: Evolution by process, not by consequence. In M.-W. Ho & S. W. Fox (Eds.), *Evolutionary processes and metaphors* (pp. 117–144). Chichester, NY: Wiley.

Ho, M.-W. (1988b). Genetic fitness and natural selection: Myth or metaphor. In E. Tobach and G. Greenberg (Eds.), *Evolution of social behavior and integra tive levels* (pp. 85–111). Hillsdale, NJ: Erlbaum.

Ho, M.-W., & Fox, S. W. (Eds.) (1988). *Evolutionary processes and metaphors.* Chichester, NY: Wiley.

Ho, M.-W., & Saunders, P. T. (1979). Beyond neo-Darwinism—an epigenetic approach to evolution. *Journal of Theoretical Biology, 78,* 573–591.

Ho, M.-W., & Saunders, P. T. (1982). Adaptation and natural selection: Mechanism and teleology. In S. Rose (Ed.), *Towards a liberatory biology* (pp. 85–102). London: Allison and Busby.

Hodges, A. (1983). *Alan Turing: The enigma.* New York: Simon and Schuster/Touchstone.

Hofer, M. A. (1981a). Parental contributions to the development of their offspring. In D. J. Gubernick & P. H. Klopfer (Eds.), *Parental care in mammals* (pp. 77–115). New York: Plenum.

Hofer, M. A. (1981b). *The roots of human behavior.* San Francisco: W. H. Freeman.

Hoffman, M. L. (1981). Is altruism part of human nature? *Journal of Personality and Social Psychology, 40,* 121–137.

Horowitz, F. D. (1969). Learning, developmental research, and individual differences. In L. P. Lipsitt & H. W. Reese (Eds.), *Advances in child development and behavior, 4,* 83–126. New York: Academic Press.

Hrdy, S. B., & Hausfater, G. (1984). Comparative and evolutionary perspectives on infanticide: Introduction and overview. In G. Hausfater and S. B. Hrdy (Eds.), *Infanticide: Comparative and evolutionary perspectives* (pp. xiii–xxxv). New York: Aldine.

Hull, D. L. (1976). Are species really individuals? *Systematic Zoology, 25,* 174–191.

Hull, D. L. (1988). *Science as a process.* Chicago: University of Chicago Press.

Immelmann, K., Barlow, G. W., Petrinovich, L., & Main, M. (1981). *Behavioral development: The Bielefeld interdisciplinary project* (pp. 1–18). Cambridge: Cambridge University Press.

Ingold, T. (1991). Becoming persons: Consciousness and sociality in human evolution. *Cultural Dynamics, 4,* 355–378.

Ingold, T. (1995). "People like us": The concept of the anatomically modern human. *Cultural Dynamics, 7,* 187–214.

Ingold, T. (1996). Life beyond the edge of nature? or, the mirage of society. In J. D. Greenwood (Ed.), *The mark of the social* (pp. 231–252). Latham, MD: Rowman and Littlefield.

Jablonka, E., & Lamb, M. J. (1995). *Epigenetic inheritance and evolution: The Lamarckian dimension.* Oxford: Oxford University Press.

Jackson, W. (1987). *Altars of unhewn stone.* San Francisco: North Point Press.

Jacobson, J. L., Boersma, D. C., Fields, R. B., & Olson, K. L. (1983). Paralinguistic features of adult speech to infants and small children. *Child Development, 54,* 436–442.

Jaffe, L. F., & Stern, C. D. (1979). Strong electrical currents leave the primitive streak of chick embryos. *Science, 206,* 569–571.

Jaggar, A. M. (1983). *Feminist politics and human nature.* Totowa, NJ: Rowman and Allanheld.

Jerne, N. K. (1967). Antibodies and learning: Selection versus instruction. In G. C. Quarton, T. Melnechuk & F. O. Schmitt (Eds.), *The neurosciences: A study program* (pp. 200–205). New York: Rockefeller University Press.

Johnson, M. (1987). *The body in the mind: The bodily basis of meaning, imagination, and reason.* Chicago: University of Chicago Press.

Johnston, T. D. (1982). Learning and the evolution of developmental systems. In H. C. Plotkin (Ed.), *Learning, development, and culture* (pp. 411–442). New York: Wiley.

Johnston, T. D. (1987). The persistence of dichotomies in the study of behavioral development. *Developmental Review, 7,* 149–182.

Johnston, T. D. (1988). Developmental explanation and the ontogeny of birdsong: Nature-nurture redux. *Behavioral and Brain Sciences, 11,* 617–663.

Johnston, T. D. (1995). The influence of Weismann's germ-plasm theory on the distinction between learned and innate behavior. *Journal of the History of the Behavioral Sciences, 31,* 115–128.

Johnston, T. D., & Gottlieb, G. (1985). Effects of social experience on visually imprinted maternal preferences in Peking ducklings. *Developmental Psychobiology, 18,* 261–271.

Johnston, T. D., & Gottlieb, G. (1990). Neophenogenesis: A developmental theory of phenotypic evolution. *Journal of Theoretical Biology, 147,* 471–495.

Johnston, T. D., & Turvey, M. T. (1980). A sketch of an ecological metatheory for theories of learning. In G. H. Bower (Ed.), *The psychology of learning and motivation* (Vol. 14) (pp. 147–205). New York: Academic Press.

Jordanova, L. J. (1984). *Lamarck.* Oxford: Oxford University Press.

Kagan, J. (1984). *The nature of the child.* New York: Basic Books.

Kauffman, S. A. (1985). Self organization, selective adaptation, and its limits: A new pattern of inference in evolution and development. In D. J. Depew & B. H. Weber (Eds.), *Evolutionary theory at the crossroads: The new bi-*

ology and the new philosophy of science (pp. 169–207). Cambridge: MIT Press/Bradford.

Kaye, K. (1982). *The mental and social life of babies.* Chicago: University of Chicago Press.

Keller, E. F. (1985). *Reflections on gender and science.* New Haven: Yale University Press.

Keller, E. F. (1987). Reproduction and the central project of evolutionary theory. *Biology and Philosophy, 2,* 73–86.

Keller, E. F. (1992). *Secrets of life, secrets of death.* New York: Routledge.

Keller, E. F., & Lloyd, E. A. (Eds.) (1992). *Keywords in evolutionary biology.* Cambridge: Harvard University Press.

Kessen, W. (1979). The American child and other cultural inventions. *American Psychologist, 34,* 815–820.

Kimura, M. (1992). Neutralism. In E. F. Keller & E. A. Lloyd, *Keywords in Evolutionary Biology* (pp. 225–230). Cambridge: Harvard University Press.

Kitcher, P. (1985). *Vaulting ambition: Sociobiology and the quest for human nature.* Cambridge: MIT Press.

Kitzinger, C. (1987). *The social construction of lesbianism.* Newbury Park, CA: Sage.

Klama, J. (a pseudonym for Durant, J., Klopfer, P. H., and Oyama, S., editors of a book cooperatively authored by Durant, J., Honore, E., Klopfer, L., Klopfer, M., Klopfer, P. H., Kohn, T., Lessley, B., Nur, N., and Oyama, S.). (1988). *The myth of the beast within: Aggression revisited.* London: Longman. (Published in the United States by Wiley as *Aggression: The myth of the beast within.*)

Klopfer, P. [H.] (1969). Instincts and chromosomes: What is an "innate" act? *American Naturalist, 103,* 556–560.

Klopfer, P. H. (1973). *On behavior: Instinct is a Cheshire cat.* Philadelphia: Lippincott.

Kohn, A. (1990). *The brighter side of human nature.* New York: Basic Books.

Konner, M. (1982). *The tangled wing: Biological constraints on the human spirit.* New York: Holt, Rinehart and Winston.

Kuo, Z.-Y. (1922). How are our instincts acquired? *Psychological Review, 29,* 344–365.

Kuo, Z.-Y. (1967/1976). *Dynamics of behavior development.* (Enlarged ed.). New York: Plenum.

Lakoff, G., & Johnson, M. (1980). *Metaphors we live by.* Chicago: University of Chicago Press.

Lamb, M. E., Pleck, J. H., Charnov, E. L., & Levine, J. A. (1985). Paternal behavior in humans. *American Zoologist, 25,* 883–894.

Latour, B. (1987). *Science in action: How to follow scientists and engineers through society.* Cambridge: Harvard University Press.

Latour, B. (1991). Technology is society made durable. In J. Law (Ed.), *A sociology of monsters: Essays on power, technology, and domination, Sociological Review Monograph 38* (pp. 103–131). London: Routledge & Kegan Paul.

Latour, B. (1993). *We have never been modern.* (Catherine Porter, Trans.). Cambridge: Harvard University Press. (Original work published in 1991.)

Lave, J. (1986). The values of quantification. In J. Law (Ed.), *Power, action, and belief: A new sociology of knowledge? Sociological Review Monograph 32* (pp. 88–111). London: Routledge & Kegan Paul.

Le Guin, U. K. (1987, September 28). Half past four. *New Yorker,* pp. 34–56.

Lehrman, D. S. (1953). A critique of Konrad Lorenz's theory of instinctive behavior. *Quarterly Review of Biology, 28,* 337–363.

Lehrman, D. S. (1962). Interaction of hormonal and experiential influences on development of behavior. In E. L. Bliss (Ed.), *Roots of behavior* (pp. 142–156). New York: Harper & Brothers.

Lehrman, D. S. (1970). Semantic and conceptual issues in the nature-nurture problem. In L. R. Aronson, E. Tobach, D. S. Lehrman & J. S. Rosenblatt (Eds.), *Development and evolution of behavior: Essays in memory of T. C. Schneirla* (pp. 17–52). San Francisco: W. H. Freeman.

Lerner, I. M., & Libby, W. J. (1976). *Heredity, evolution, and society.* (2nd ed.). San Francisco: W. H. Freeman.

LeVay, S. (1993). *The sexual brain.* Cambridge: MIT Press.

LeVay, S. (1996). *Queer science: The use and abuse of research into homosexuality.* Cambridge, MA: MIT Press.

Levins, R. (1981, Winter). Class science & scientific truth. *Working Papers on Marxism & Science, 1,* 9–22.

Levins, R., & Lewontin, R. (1985). *The dialectical biologist.* Cambridge: Harvard University Press.

Lewin, R. (1984). Why is development so illogical? *Science, 224,* 1327–1329.

Lewontin, R. C. (1970/1976). Race and intelligence. In N. J. Block & G. Dworkin (Eds.), *The IQ controversy* (pp. 78–92). New York: Pantheon. (Reprinted from *Bulletin of the Atomic Scientists,* pp. 2–8, March, 1970.)

Lewontin, R. (1978, September). Adaptation. *Scientific American, 239,* 212–230.

Lewontin, R. C. (1982). Organism and environment. In H. C. Plotkin (Ed.), *Learning, development, and culture* (pp. 151–170). New York: Wiley.

Lewontin, R. C. (1983a, June 16). Darwin's revolution. *New York Review of Books, 30*(10), 21–27.

Lewontin, R. C. (1983b). Gene, organism, and environment. In D. S. Bendall (Ed.), *Evolution from molecules to men* (pp. 273–285). Cambridge: Cambridge University Press.

Lewontin, R. C. (1984). *Human diversity.* San Francisco: W. H. Freeman.

Lewontin, R. C. (1974/1985). The analysis of variance and the analysis of causes. In Levins, R., & Lewontin, R., *The dialectical biologist* (pp. 109–122). Cambridge: Harvard University Press. (Reprinted from *American Journal of Human Genetics, 26,* 400–411, 1974.)

Lewontin, R. C. (1992). Genotype and phenotype. In E. F. Keller & E. A. Lloyd (Eds.), *Keywords in evolutionary biology* (pp. 137–144). Cambridge: Harvard University Press.

Lewontin, R. C., Rose, S., & Kamin, L. J. (1984). *Not in our genes.* New York: Pantheon.

Lickliter, R., & Gottlieb, G. (1985). Social interaction with siblings is necessary for visual imprinting of species-specific maternal preferences in ducklings. *Journal of Comparative Psychology, 99,* 371–379.

Lickliter, R., & Gottlieb, G. (1988). Social specificity: Interaction with own species is necessary to foster species-specific maternal preference in ducklings. *Developmental Psychobiology, 21,* 311–321.

Lindzey, G., Hall, C. S., & Thompson, R. F. (1978). *Psychology.* (2nd ed.). New York: Worth.

Lockard R. B. (1971). Reflections on the fall of comparative psychology: Is there a message for us all? *American Psychologist, 26,* 168–179.

Longino, H. (1990). *Science as social knowledge.* Princeton: Princeton University Press.

Longino, H., & Doell, R. (1987). Body, bias, and behavior: A comparative analysis of reasoning in two areas of biological science. In S. Harding & J. F. O'Barr (Eds.), *Sex and scientific inquiry* (pp. 165–186). Chicago: University of Chicago Press. (Reprinted from *Signs, 9*(1), 1983.)

Lorenz, K. (1965). *Evolution and modification of behavior.* Chicago: University of Chicago Press.

Lorenz, K. (1973/1977). *Behind the mirror.* (R. Taylor, Trans.). New York: Harcourt, Brace and Jovanovich. (Original work published in 1973.)

Lumsden, C. J., & Wilson, E. O. (1981). *Genes, mind, and culture.* Cambridge: Harvard University Press.

MacDonald, K. B. (Ed.). (1988a). *Sociobiological perspectives on human development.* New York: Springer-Verlag.

MacDonald, K. B. (1988b). Sociobiology and the cognitive-developmental tradition in moral development research. In K. B. MacDonald (Ed.), *Sociobiological perspectives on human development* (pp. 140–167). New York: Springer-Verlag.

Magnusson, D., & Allen, V. L. (Eds.). (1983). *Human development: An interactional perspective.* New York: Academic Press.

Margolis, H. (1987). *Selfishness, altruism, and rationality: A theory of social choice.* Chicago: University of Chicago Press.

Margulis, L. (1981). *Symbiosis in cell evolution: Life and its environment on the early Earth.* San Francisco: W. H. Freeman.

Markert, C. L., & Ursprung, H. (1971). *Developmental genetics.* Englewood Cliffs, NJ: Prentice-Hall.

Masters, R. D. (1986). Ostracism, voice, and exit: The biology of social participation. *Ethology and Sociobiology, 7,* 379–395.

Maturana, H., & Varela, F. (1987). *The tree of knowledge.* Boston: New Science Library.

Maynard Smith, J. (1978/1984). Optimization theory in evolution. In E. Sober (Ed.), *Conceptual issues in evolutionary biology* (pp. 289–314). Cambridge: MIT Press/Bradford. (Reprinted from *Annual Review of Ecology and Systematics, 9,* 331–56, 1978.)

Maynard Smith, J. (1986). *The problems of biology.* Oxford: Oxford University Press.

Maynard Smith, J., Burian, R., Kauffman, S., Alberch, P., Campbell, H., Goodwin, B., Lande, R., Raup, D., & Wolpert, L. (1985). Developmental constraints and evolution. *Quarterly Review of Biology, 60,* 265–287.

Mayr, E. (1961). Cause and effect in biology. *Science, 134,* 1501–1506.

Mayr, E. (1976a). Behavior programs and evolutionary strategies. In E. Mayr (Ed.), *Evolution and the diversity of life* (pp. 694–711). Cambridge: Harvard University Press/Belknap Press.

Mayr, E. (1976b). Cause and effect in biology. In E. Mayr (Ed.), *Evolution and the diversity of life* (pp. 359–371). Cambridge: Harvard University Press/Belknap Press.

Mayr, E. (1982). *The growth of biological thought.* Cambridge: Harvard University Press/Belknap Press.

McAllister, P. (Ed.). (1982). *Reweaving the web of life.* Philadelphia: New Society.

McClearn, G. E., & DeFries, J. C. (1973). *Introduction to behavioral genetics.* San Francisco: W. H. Freeman.

Meadows, D. (1988). World interconnectedness also works in our favor. *Annals of Earth 6*(1), 16.

Midgley, M. (1980). *Beast and man: The roots of human nature.* New York: New American Library.

Milkman, R. (Ed.). (1982). *Perspectives on evolution.* New York: Sinauer.

Money, J., & Ehrhardt, A. A. (1972). *Man and woman, boy and girl.* Baltimore: Johns Hopkins University Press.

Monod, J. (1971). *Chance and necessity.* (A. Wainhouse, Trans.). New York: Knopf.

Moore, C. L. (1984). Maternal contributions to the development of masculine sexual behavior in laboratory rats. *Developmental Psychobiology, 17,* 347–356.

Morris, E. K. (1988). Contextualism: The world view of behavior analysis. *Journal of Experimental Child Psychology, 46,* 289–323.

Morss, J. (1990). *The biologising of childhood: Developmental psychology and the Darwinian myth.* Hillsdale, NJ: Erlbaum.

Moss, L. (1992). A kernel of truth? On the reality of the genetic program. In D. L. Hull, M. Forbes & K. Okruhlik (Eds.), *Philosophy of Science Association Proceedings, 1,* 335–348.

Munn, N. L. (1965). *The evolution and growth of human behavior.* (2nd ed.). Boston: Houghton Mifflin.

Nash, J. (1970). *Developmental psychology: A psychobiological approach.* Englewood Cliffs, NJ: Prentice-Hall.

Nelkin, D., & Lindee, M. S., (1995). *The DNA mystique: The gene as a cultural icon.* New York: W. H. Freeman.

Neumann-Held, E. (1999). The gene is dead—long live the gene. In P. Koslowski (Ed.), *Sociobiology and bioeconomics: The theory of evolution in biological and economic theory* (pp. 105–137). Berlin: Springer-Verlag.

Netting, R. M. (1978). *Cultural ecology.* Menlo Park, CA: Cummings.

Newman, S. A. (1988). Idealist biology. *Perspectives in Biology and Medicine, 31,* 353–368.

Nijhout, H. F. (1990). Metaphors and the role of genes in development. *BioEssays, 12,* 441–446.

Nitecki, M. H. (Ed). (1983). *Coevolution.* Chicago: University of Chicago Press.

Nitecki, M. H. (Ed.). (1989). *Evolutionary progress.* Chicago: University of Chicago Press.

Noonan, K. M. (1987). Evolution: A primer for psychologists. In C. B. Crawford, M. S. Smith, & D. Krebs (Eds.). *Sociobiology and psychology* (pp. 31–60). Hillsdale, NJ: Erlbaum.

Oppenheim, R. W. (1980). Metamorphosis and adaptation in the behavior of developing organisms. *Developmental Psychobiology, 13,* 353–356.

Oyama, S, (1979). The concept of the sensitive period in developmental studies. *Merrill-Palmer Quarterly, 25,* 83–103.

Oyama, S. (1982). A reformulation of the idea of maturation. In P. P. G. Bateson & P. H. Klopfer (Eds.), *Perspectives in ethology* (Vol. 5) (pp. 101–131). New York: Plenum.

Oyama, S. (1987). Looking for nature. Presentation at the Fifteenth Anniversary Meeting of the Lindisfarne Association, New York City, November 7.

Oyama, S. (1988). Populations and phenotypes. [Review of the book, *Develop-*

ment, genetics, and psychology, and reply to Plomin.] *Developmental Psychobiology, 21,* 97–100, 101–105.

Oyama, S. (1989). Innate selfishness, innate sociality. *Behavioral and Brain Sciences, 12,* 717–718.

Oyama, S. (1990a). Commentary: The idea of innateness: Effects on language and communication research. *Developmental Psychobiology, 23,* 741–747.

Oyama, S. (1990b). [Review of the book, *Biology and freedom.*] *International Journal of Comparative Psychology, 3,* 191–194.

Oyama, S. (1992). Pensare d'evoluzione. L'integrazione del contesto nell'ontogenesi, nella filogenesi, nella cognizione (Thinking about evolution: Integrating the context in ontogeny, phylogeny, and cognition). In M. Ceruti (Ed.), *Evoluzione e cognizione. L'epistemologia genetica di Jean Piaget e le prospettive del costruttivismo* (pp. 47–60). Bergamo: Lubrina Editore. (Published in French, 1993, *Intellectica 16*(1), 133–150.)

Oyama, S. (1993). Constraints and development. *Netherlands Journal of Zoology, 43*(1–2), 6–16.

Oyama, S. (1994). Rethinking development. In P. Bock (Ed.), *Handbook of psychological anthropology* (pp. 185–196). Westport: Greenwood.

Oyama, S. (1999). Locating development, locating developmental systems. In E. K. Scholnick, K. Nelson, S. A. Gelman, & P. H. Miller (Eds.), *Conceptual development: Piaget's legacy* (pp. 185–208). Hillsdale, NJ: Erlbaum.

Oyama, S. (2000). *The ontogeny of information: Developmental systems and evolution.* (2nd ed.). Durham: Duke University Press.

Oyama, S., Dunbar, R. I. M., Goodwin, B. C., Strathern, M., & Ingold, T. (1996). Discussion, *Cultural Dynamics, 8,* 353–386.

Oyama, S., Griffiths, P. E., & Gray, R. D. (Eds.) (in press). *Cycles of contingency: Developmental systems and evolution.* Cambridge: MIT Press.

Patten, B. C. (1982). Environs: Relativistic elementary particles for ecology. *American Naturalist, 119,* 179–219.

Peterson, S. A., & Somit, A. (1978). Sociobiology and politics. In A. L. Caplan (Ed.), *The sociobiology debate* (pp. 449–461). New York: Harper & Row.

Piaget, J. (1978). *Behavior and evolution* (D. Nicholson-Smith, Trans.). New York: Pantheon. (Original work published in 1976.)

Piattelli-Palmarini, M. (1989). Evolution, selection and cognition: From "learning" to parameter setting in biology and in the study of language. *Cognition, 31,* 1–44.

Pinker, S. (1994). *The language instinct.* New York: William Morrow.

Pinker, S. (1997). *How the mind works.* New York: Norton.

Plagens, P. (1990, August 20). Palms and circumstance. *Newsweek,* p. 64.

Plomin, R. (1986). *Development, genetics, and psychology.* Hillsdale, NJ: Erlbaum.

Plomin, R., & Rowe, D. C. (1978). Genes, environment, and development of temperament in young human twins. In G. M. Burghardt & M. Bekoff (Eds.), *The development of behavior: Comparative and evolutionary aspects* (pp. 279–296). New York: Garland Press.

Raff, R. A., & Kaufman, T. C. (1983). *Embryos, genes, and evolution: The developmental-genetic basis of evolutionary change.* New York: Macmillan.

Rappoport, L. (1986). Renaming the world: On psychology and the decline of positive science. In S. Larsen (Ed.), *Dialectics and ideology in psychology* (pp. 167–195). Norwood, NJ: Ablex.

Riley, D. (1978). Developmental psychology: Biology and Marxism. *Ideology & Consciousness, 4,* 73–91.

Rodgers, D. A. (1970). Mechanism-specific behavior: An experimental alternative. In G. Lindzey & D. D. Thiessen (Eds.), *Contributions to behavior-genetic analysis* (pp. 207–218). New York: Appleton-Century-Crofts.

Rogoff, B., & Wertsch, J. V. (Eds.). (1984). *Children's learning in the "zone of proximal development."* San Francisco: Jossey-Bass.

Rose, N. (1985). *The psychological complex.* London: Routledge & Kegan Paul.

Rosenthal, P. (1984). *Words & values.* Oxford: Oxford University Press.

Salamone, C. (1982). The prevalence of the Natural Law within women: Women and animal rights, In P. McAllister (Ed.), *Reweaving the web of life* (pp. 364–375). Philadelphia: New Society.

Salthe, S. N. (1985). *Evolving hierarchical systems.* New York: Columbia University Press.

Sampson, E. E. (1991). *Social worlds, personal lives.* San Diego: Harcourt Brace Jovanovich.

Sapp, J. (1987). *Beyond the gene: Cytoplasmic inheritance and the struggle for authority in genetics.* Oxford: Oxford University Press.

Sayers, J. (1982). *Biological politics.* London: Tavistock.

Scarr, S. (1981). Genetics and the development of intelligence. In S. Scarr (Ed.), *Race, social class, and individual differences in I.Q.* (pp. 3–59). Hillsdale, NJ: Erlbaum.

Scarr, S., & McCartney, K. (1983). How people make their own environments: A theory of genotype → environment effects. *Child Development, 54,* 424–435.

Schaffner, K. F. (1998). Genes, behavior, and developmental emergentism: One process, indivisible? *Philosophy of Science, 65,* 209–252.

Schneirla, T. C. (1966). Behavioral development and comparative psychology. *Quarterly Review of Biology, 41,* 283–302.

Schneirla, T. C. (1956/1972). Interrelationships of the "innate" and the "acquired" in instinctive behavior. In Schneirla, T. C., *Selected writings* (pp. 131–188). San Francisco: W. H. Freeman. (Reprinted from P. P. Grassé, (Ed.),

1956, *L'Instinct dans le comportement des animaux et de l'homme.* Paris: Masson et Cie.)

Schwartz, B. (1986). *The battle for human nature: Science, morality, and modern life.* New York: Norton.

Searle, J. R. (1980). Minds, brains, and programs. *Behavioral and Brain Sciences, 3,* 417–457.

Shanahan, T. (1997). Kitcher's compromise: A critical evaluation of the compromise model of scientific closure, and its implications for the relationship between history and philosophy of science. *Studies in History and Philosophy of Science, 28,* 319–338.

Shapin, S. (1994). *A social history of truth.* Chicago: University of Chicago Press.

Shatz, M., (1985). An evolutionary perspective on plasticity in language development: A commentary. *Merrill-Palmer Quarterly, 31,* 211–222.

Shaver, K. G. (1985). *The attribution of blame.* New York: Springer-Verlag.

Shettleworth, S. J. (1972). Constraints on learning. *Advances in the Study of Behavior, 4,* 1–68.

Shields, W. M., & Shields, L. M. (1983). Forcible rape: An evolutionary perspective. *Ethology and Sociobiology, 7,* 115–136.

Shotter, J. (1984). *Social accountability and selfhood.* Oxford: Blackwell.

Shotter, J. (1993). *Cultural politics of everyday life: Social constructionism, rhetoric, and knowing of the third kind.* Toronto: University of Toronto Press.

Shweder, R. A. (1990). Cultural psychology—what is it? In J. W. Stigler, R. A. Shweder, & G. Herdt (Eds.), *Cultural psychology: Essays on comparative human development* (pp. 1–43). Cambridge: Cambridge University Press.

Silverman, I. (1987). Race, race differences, and race relations: Perspectives from psychology and sociobiology. In C. B. Crawford, M. S. Smith, & D. Krebs (Eds.). *Sociobiology and psychology* (pp. 205–221). Hillsdale, NJ: Erlbaum.

Skinner, B. F. (1971). *Beyond freedom and dignity.* New York: Bantam/Vintage.

Skinner, B. F. (1981). Selection by consequence. *Science, 213,* 501–504.

Smith, B. H. (1988). *Contingencies of value: Alternative perspectives for critical theory.* Cambridge: Harvard University Press.

Smith, B. H. (1991). Belief and resistance: A symmetrical account. *Critical Inquiry, 18*(1), 125–139.

Smith, B. H. (1992). The unquiet judge: Activism without objectivism in law and politics. *Annals of Scholarship, 9,* 111–133.

Smith, B. H. (1997). *Belief and resistance: Dynamics of contemporary intellectual controversy.* Cambridge: Harvard University Press.

Smith, B. H. (1998, February 20). Is it really a computer? [Review of the book, *How the mind works*]. *Times Literary Supplement,* pp. 3–4.

Smith, B. H. (in press). Cutting-edge equivocation: Conceptual moves and rhetorical strategies in contemporary anti-epistemology. *South Atlantic Quarterly.*

Smith, B. H., & Plotnitsky, A. (Eds.) (1997). *Mathematics, science, and postclassical theory.* Durham: Duke University Press.

Smith, K. C. (1994). *The emperor's new genes: The role of the genome in development and evolution.* Ph.D. dissertation, Duke University.

Smith, M. S. (1987). Evolution and developmental psychology: Toward a sociobiology of human development. In C. B. Crawford, M. S. Smith, & D. Krebs (Eds.), *Sociobiology and psychology* (pp. 225–252). Hillsdale, NJ: Erlbaum.

Sober, E. (1980). Evolution, population thinking, and essentialism. *Philosophy of Science, 47,* 350–383.

Sober, E. (1984). *The nature of selection.* Cambridge: MIT Press/Bradford.

Sober, E. (1985). Darwin on natural selection: A philosophical perspective. In D. Kohn (Ed.), *The Darwinian heritage* (pp. 867–899). Princeton: Princeton University Press.

Sober, E. (1987). What is adaptationism? In J. Dupre (Ed.), *The latest on the best* (pp. 105–118). Cambridge: MIT Press.

Sober, E. (1988). What is evolutionary altruism? *Canadian Journal of Philosophy, 14,* 75–99.

Sober, E. (1992). Models of cultural evolution. In P. E. Griffiths (Ed.), *Trees of life: Essays in philosophy of biology* (pp. 17–39). Boston: Kluwer.

Sober, E., & Wilson, D. S. (1998). *Unto others: The evolution and psychology of unselfish behavior.* Cambridge: Harvard University Press.

Star, S. L. (1991). Power, technology, and the phenomenology of conventions: On being allergic to onions. In J. Law (Ed.), *A sociology of monsters: Essays on power, technology, and domination, Sociological Review Monograph 38* (pp. 26–56). London: Routledge & Kegan Paul.

Stearns, S. C. (1986). Natural selection and fitness, adaptation and constraint. In D. M. Raup & D. Jablonski (Eds.), *Pattern and process in the history of life* (pp. 23–44). Berlin: Springer-Verlag.

Stent, G. S. (1981). Strength and weakness of the genetic approach to the development of the nervous system. In W. M. Cowen (Ed.), *Studies in developmental neurobiology* (pp. 288–321). Oxford: Oxford University Press. (Adapted from *Annual Review of Neuroscience, 4,* 163–193.)

Sterelny, K. (1995). [Review of the book, *The adapted mind.*] *Biology and Philosophy, 10,* 365–380.

Sterelny, K., & Griffiths, P. E. (1999). *Sex and death: An introduction to philosophy of biology.* Chicago: University of Chicago Press.

Sterelny, K., Smith, K. C., & Dickison, M. (1996). The extended replicator, *Biology and Philosophy, 11,* 377–403.

Stern, K. (1973). *Principles of human genetics.* (3rd ed.). San Francisco: W. H. Freeman.

Strohman, R. C. (1997, March). The coming Kuhnian revolution in biology. *Nature Biotechnology, 15,* 194–200.

Studdert-Kennedy, M. (1990). Language development from an evolutionary perspective. *Haskins Laboratories Status Report on Speech Research, SR-1101/102,* 14–27.

Symons, D. (1987). If we're all Darwinians, what's the fuss about? In C. B. Crawford, M. S. Smith, & D. Krebs (Eds.). *Sociobiology and psychology* (pp. 121–146). Hillsdale, NJ: Erlbaum.

Taylor, P. J. (1987). Historical versus selectionist explanations in evolutionary biology, *Cladistics, 3,* 1–13.

Taylor, P. J. (1988). Technocratic optimism and the partial transformation of ecological metaphor after World War Two. *Journal of the History of Biology 21*(2), 213–244.

Taylor, P. J. (1992). Community. In E. F. Keller & E. A. Lloyd, *Keywords in evolutionary biology* (pp. 52–60). Cambridge: Harvard University Press.

Taylor, P. J. (1995). Building on construction: An exploration of heterogeneous constructionism, using an analogy from psychology and a sketch from socio-economic modeling. *Perspectives on Science, 3,* 66–98.

Thelen, E., & Smith, L. B. (1994). *A dynamic systems approach to the development of cognition and action.* Cambridge: MIT Press/Bradford.

Thiessen, D. D. (1972). *Gene organization and behavior.* New York: Random House.

Thomson, K. (1985). Essay review: The relationship between development and evolution. *Oxford Surveys in Evolutionary Biology, 2,* 220–233.

Thornhill, R., & Thornhill, N. M. (1987). Human rape: The strengths of the evolutionary perspective. In C. B. Crawford, M. S. Smith, & D. Krebs (Eds.), *Sociobiology and psychology* (pp. 269–291). Hillsdale, NJ: Erlbaum.

Tiger, L. (1970). *Men in groups.* New York: Vintage Books.

Tiger, L., & Fox, R. (1971). *The imperial animal.* New York: Holt, Rinehart and Winston.

Tinbergen, N. (1963). On aims and methods in ethology. *Zeitschrift für Tierpsychologie, 20,* 410–433.

Tobach, E. (1972). The meaning of the cryptanthroparion. In L. Ehrman, G. S. Omenn, & E. Caspari (Eds.), *Genetics, environment, and behavior* (pp. 219–239). New York: Academic Press.

Tobach, E., & Greenberg, G. (1984). The significance of T. C. Schneirla's contribution to the concept of levels of integration. In G. Greenberg & E. Tobach (Eds.), *Behavioral evolution and integrative levels* (pp. 1–7). Hillsdale, NJ: Erlbaum.

Tooby, J., & Cosmides, L. (1992). The psychological foundations of culture. In J. H. Barkow, L. Cosmides, & J. Tooby (Eds.), *The adapted mind* (pp. 19–136). Oxford: Oxford University Press.

Topoff, H. (1974). Genes, intelligence, and race. In E. Tobach, G. Gianutsos, H. R. Topoff, & C. G. Gross (Eds.), *The four horsemen: Racism, sexism, militarism, and social Darwinism* (pp. 25–66). New York: Behavioral Publications.

Trevarthen, C. (1982). The primary motives for cooperative understanding. In G. Butterworth & P. Light (Eds.), *Social cognition: Studies of the development of understanding* (pp. 77–109). Brighton: Harvester Press.

Turing, A. M. (1950/1964). Computing machinery and intelligence. In A. R. Anderson (Ed.), *Minds and machines* (pp. 4–30). Englewood Cliffs, NJ: Prentice-Hall. (Reprinted from *Mind, 59,* 433–460, 1950.)

Turkewitz, G., & Kenny, P. A. (1982). Limitations on input as a basis for neural organization and perceptual development: A preliminary theoretical statement. *Developmental Psychobiology, 15,* 357–368.

Valsiner, J. (1987). *Culture and the development of children's action.* New York: Wiley.

Valsiner, J. (1991). Construction of the mental. *Theory & Psychology, 1,* 477–494.

Vandenbergh, J. G. (1987). Regulation of puberty and its consequences on population dynamics of mice. *American Zoologist, 27,* 891–898.

van den Berghe, P. L. (1987). Incest taboos and avoidance: Some African applications. In C. B. Crawford, M. S. Smith, & D. Krebs (Eds.). *Sociobiology and psychology* (pp. 353–371). Hillsdale, NJ: Erlbaum.

van der Weele, C. (1993). Explaining embryological development: Should integration be the goal? *Biology and Philosophy, 8,* 385–397.

van der Weele, C. (1999). *Images of development: Environmental causes in ontogeny.* Albany, NY: SUNY Press.

van Gelder, T., & Port, R. F. (1995). It's about time: An overview of the dynamical approach to cognition. In R. F. Port & T. van Gelder (Eds.), *Mind as motion: Explorations in the dynamics of cognition* (pp. 1–43). Cambridge: MIT Press/Bradford.

Varela, F. J., Thompson, E., & Rosch, E. (1991). *The embodied mind.* Cambridge: MIT Press.

Venn, C. (1984). The subject of psychology. In J. Henriques, W. Hollway, C. Urwin, C. Venn, & V. Walkerdine, *Changing the subject* (pp. 119–152). London: Methuen.

Voorzanger, B. (1987a). No norms and no nature—the moral relevance of evolutionary biology. *Biology and Philosophy, 2,* 39–56.

Voorzanger, B. (1987b). Methodological problems in evolutionary biology VIII. Biology and culture. *Acta Biotheoretica, 36,* 23–34.

Waddington, C. H. (1957). *The strategy of the genes.* London: George Allen and Unwin.

Waddington, C. H. (1975). *The evolution of an evolutionist.* Ithaca: Cornell University Press.

Weber, B. H., & Depew, D. J. (1999). Does the Second Law of Thermodynamics refute the Neo-Darwinian synthesis? In P. Koslowski (Ed.), *Sociobiology and bioeconomics: The theory of evolution in biological and economic theory* (pp. 50–75). Berlin: Springer-Verlag.

Webster, G., & Goodwin, B. C. (1982). The origin of species: A structuralist approach. *Journal of Social and Biological Structures, 5,* 15–47.

Weismann, A. (1893). *The germ plasm: A theory of heredity.* New York: Scribner's.

Weiss, P. A. (1969). The living system: Determinism stratified. In A. Koestler & J. R. Smythies (Eds.), *Beyond reductionism* (pp. 3–55). New York: Macmillan.

West, M. J., & King, A. P. (1987). Settling nature and nurture into an ontogenetic niche. *Developmental Psychobiology, 20,* 549–562.

West, M. J., & King, A. P. (1988). Female visual displays affect the development of male song in the cowbird. *Nature, 334,* 244–246.

Wexler, K., & Culicover, P. (1980). *Formal principles of language acquisition.* Cambridge, MA: MIT Press.

Whelan, R. E. (1971). The concept of instinct. In J. L. McGaugh (Ed.), *Psychobiology: Behavior from a biological perspective* (pp. 53–72). New York: Academic Press.

Williams, G. C. (1966). *Adaptation and natural selection.* Princeton: Princeton University Press.

Williams, G. C. (1985). Comments by George C. Williams on Sober's *The nature of selection. Biology and Philosophy, 1,* 114–122.

Wilson, D. S. (1989). Levels of selection: An alternative to individualism in biology and the human sciences. *Social Networks, 11,* 257–272.

Wilson, E. O. (1978). *On human nature.* Cambridge: Harvard University Press.

Wimsatt, W. C. (1980/1984). Reductionistic research strategies and their biases in the units of selection controversy. In E. Sober (Ed.), *Conceptual issues in evolutionary biology* (pp. 142–183). Cambridge: MIT Press/Bradford. (Reprinted from T. Nickles, Ed., 1980, *Scientific discovery,* Vol. 2: *Historical and scientific case studies,* pp. 213–259, Dordrecht: D. Reidel.)

Wimsatt, W. C. (1986). Developmental constraints, generative entrenchment, and the innate-acquired distinction. In W. Bechtel (Ed.), *Integrating scientific disciplines* (pp. 185–208). Dordrecht: Martinus-Nijhoff.

Winther, R. (1996). *Constructions of form in biological systems.* Master's thesis, Stanford University.

Winther, R. (in press). Weismann on germ-plasm variation. *Journal of the History of Biology.*

Wolters, G. (1995). Comment on Brandon and Antonovics' "Coevolution of Organism and Environment." In G. Wolters & J. G. Lennow (Eds., in collaboration with P. McLaughlin), *Concepts, theories, and rationality in the biological sciences* (pp. 232–240). Pittsburgh: University of Pittsburgh Press.

Young, R. M. (1985). *Darwin's metaphor.* Cambridge: Cambridge University Press.

Yoxen, E. (1981). Life as a productive force: Capitalising the science and technology of molecular biology. In R. M. Young & L. Levidow (Eds.), *Studies in the labour process* (pp. 66–122). London: CSE Books.

Yuwiler, A. (1971). Problems in assessing biochemical ontogeny. In M. B. Sterman, D. J. McGinty, & A. M. Adionolfi (Eds.), *Brain development and behavior* (pp. 43–57). New York: Academic Press.

Zanotti, B. (1982). Patriarchy: A state of war. In P. McAllister (Ed.), *Reweaving the web of life* (pp. 16–19). Philadelphia: New Society.

Index of Subjects and Names

Susan Oyama is Professor of Psychology at John Jay
College, CUNY. She is the author of *The Ontogeny of
Information: Developmental Systems and Evolution*
(Duke University Press, 2000).

Library of Congress Cataloging-in-Publication Data

Oyama, Susan.
Evolution's eye : A systems view of the biology-culture
divide / Susan Oyama.
p.   cm. — (Science and cultural theory)
Includes bibliographical references (p.   ) and index.
ISBN 0-8223-2436-9 (cloth : alk. paper). — ISBN
0-8223-2472-5 (paper : alk. paper)
1. Developmental psychology.   2. Genetic psychology.
3. System theory.   I. Title.   II. Series.
BF713.093   2000
155.7—dc21   99-38175
                                              CIP